Transboundary Water Politics in the Developing World

This book examines the political economy that governs the management of international transboundary river basins in the developing world. These shared rivers are the setting for irrigation, hydropower and flood management projects as well as water transfer schemes. Often, these projects attempt to engineer the river basin with deep political, socio-economic and environmental implications. The politics of transboundary river basin management sheds light on the challenges concerning sustainable development and utilization between sovereign states.

Advancing conceptual thinking beyond simplistic analyses of river basins in conflict or cooperation, the author proposes a new analytical framework. The transboundary Waters Interaction NexuS (TWINS) examines the coexistence of conflict and cooperation in riparian interaction. This framework highlights the importance of power relations between basin states that determine negotiation processes and institutions of water resources management. The analysis illustrates the way river basin management is framed by powerful elite decision-makers, combined with geopolitical factors and geographical imaginations. In addition, the book explains how national development strategies and water resources demands have a significant role in shaping the intensities of conflict and cooperation at the international level

The book draws on detailed case studies from the Ganges River basin in South Asia, the Orange-Senqu River basin in Southern Africa and Mekong River basin in Southeast Asia, providing key insights on equity and power asymmetry applicable to other basins in the developing world.

Naho Mirumachi is a Lecturer in Geography at King's College London, UK.

Earthscan studies in water resource management

Sustainable Water and Sanitation Services
The life-cycle approach to planning and management
Livelihoods & Natural Resource Management Institute, International Water & Sanitation Centre, Centre for Economic and Social Studies, Watershed Support Services & Activities Network

Water for Food Security and Well-being in Latin America and the Caribbean
Social and environmental implications for a globalized economy
Edited by Bárbara A. Willaarts, Alberto Garrido and M. Ramón Llamas

Water Scarcity, Livelihoods and Food Security
Research and innovation for development
Edited by Larry W. Harrington, Myles J. Fisher

Adaptation to Climate Change through Water Resources Management
Capacity, equity and sustainability
Edited by Dominic Stucker and Elena Lopez-Gunn

Hydropower Development in the Mekong Region
Political, socio-economic and environmental perspectives
Edited by Nathanial Matthews and Kim Geheb

Governing Transboundary Waters
Canada, the United States and Indigenous Communities
Emma S. Norman

Transboundary Water Politics in the Developing World
Naho Mirumachi

Water and Cities in Latin America
Challenges for sustainable development
Edited by Ismael Aguilar-Barajas, Jürgen Mahlknecht, Jonathan Kaledin and Marianne Kjellén

For more information and to view forthcoming titles in this series, please visit the Routledge website: www.routledge.com/books/series/ECWRM/

Transboundary Water Politics in the Developing World

Naho Mirumachi

LONDON AND NEW YORK

from Routledge

First published 2015
by Routledge
2 Park Square, Milton Park, Abingdon, Oxon OX14 4RN

and by Routledge
711 Third Avenue, New York, NY 10017

Routledge is an imprint of the Taylor & Francis Group, an informa business

British Library Cataloguing-in-Publication Data
A catalogue record for this book is available from the British Library

Library of Congress Cataloging in Publication Data
Mirumachi, Naho.
Transboundary water politics in the developing world /
Naho Mirumachi.
    pages cm. – (Earthscan studies in water resource management)
    Summary: "This book examines the political economy that governs
    the development and management of international transboundary
    river basins" – Provided by publisher.
    Includes bibliographical references and index.
    1. Water resources–Government policy–Developing countries.
    2. Water-supply–Developing countries–Management. I. Title.
    HD1702.M57 2015
    333.91009172'4–dc23   2014037786

ISBN: 978-0-415-81295-5 (hbk)
ISBN: 978-0-415-81296-2 (pbk)
ISBN: 978-0-203-06838-0 (ebk)

Typeset in Bembo
by Wearset Ltd, Boldon, Tyne and Wear

The importance of this book extends beyond transboundary water and the developing world to bring intelligent insights to many water issues. Naho Mirumachi critically examines a number of shibboleths in water management such as: conflict and cooperation are opposites; agreement means decisions are equitable; templates like the creation of strong river basin organization structures leads to better decisions; and increasing resources for the hydrocracy serves the public interest. This is the best book about water I have read in a long time.

Helen Ingram, *The Southwest Center, University of Arizona and Professor Emeritus, University of California at Irvine, USA*

Transboundary Water Politics in the Developing World brings insight and nuance to a field of inquiry dominated by speculation and generalisation. We often hear that wars of the future will be fought over water by states locked in competition for a shared but increasingly scarce commodity. But we are also told that it is far more common for states to cooperate than it is for them to conflict over transboundary waters. Applying a rich theoretical framework dubbed 'TWINS' – the transboundary waters interaction nexus – to three case studies, Naho Mirumachi sheds new light not only in terms of her cases – the Mekong, Ganges–Brahmaputra, and Orange–Senqu river basins – but in terms of what we know and what we can expect from transboundary water politics. This is a must-read for all those interested in the complex ways a multiplicity of actors, forces and factors come together in the simultaneously conflictful and cooperative world of water resources development and management

Larry A. Swatuk, *University of Waterloo, Canada*

Contents

Figures

Acknowledgements

This book is published in 2015, which happens to be the closing year of the 'UN Decade of Water for Life'. During these past ten years, I have noticed that there has been more awareness around water issues, if not for the increasing number of people attending major 'water' conferences and international fora. Yet at the same time it seems that there are recurring unanswered themes and questions, especially with regard to the governance and politics of water resources. This book is an attempt to dig deeper into some of these themes and questions.

My interest in transboundary water issues has allowed me to meet many enlightened and critical minds. My thinking has benefited from the intellectual rigour and friendships found within the London Water Research Group (LWRG). I owe much to Tony Allan as a mentor, colleague and friend. His drive for major original ideas continues to inspire me. When he kindly read a draft of the manuscript, he mentioned that I should 'write for the world'. While imperfect, this book is a first stab at pointing out what *really* matters in the politics of water allocation and management. I have enjoyed working with Mark Zeitoun, Jeroen Warner and Ana Cascão over the years. They continue to probe and challenge transboundary water issues, all the while mixing in some good fun whenever/wherever we meet. While my ideas have since evolved, Elizabeth Kistin Keller, Suvi Sojamo, Clemens Messerschmid and Francesca Greco seriously engaged with the idea of TWINS when I first presented to the LWRG.

At my academic home of the Geography Department at King's College London, the discussions with my Environment, Politics and Development (EPD) research group colleagues have helped shape my thoughts. In particular, I thank Raymond Bryant, Frezer Haile, Mike Hulme and Alex Loftus for reviewing earlier versions of my chapters. Their comments helped me to realize the power of narrative that a book can bring about – something very different from a journal paper. Being around an interdisciplinary 'water community' at King's, in addition to the Master's students on the Water: Science and Governance programme, has been motivational.

Those working within, and often across, academia and policy have provided greater insight into the river basins I have visited: Dwarika Dhungel,

Ajaya Dixit, John Dore, Kim Geheb, Dipak Gyawali, Oliver Hensengerth, Marko Keskinen, Matti Kummu, Nate Matthews, Carl Middleton, Santa Pun and Tony Turton. In addition, many interviewees kindly shared their thoughts and time during fieldwork. I am grateful to the British Academy Small Research Grants (SG112581) for enabling fieldwork to delve into the Mekong case study, and for the opportunity to further my thinking on scales and actors, crucial to the development of this book.

Through these discussions, conversations and field experiences, I have been able to sharpen my arguments and refine ideas, some of which have been presented in earlier work. Sections of this book draw on material from the following publications for revised and reworked content: Mirumachi, N. (2013) 'Securitising shared waters: An analysis of the hydropolitical context of the Tanakpur Barrage project between Nepal and India', *Geographical Journal*, vol. 179, no. 4, pp. 309–319, published by John Wiley & Sons, © 2013 The Author, the Geographical Journal © 2013 Royal Geographical Society (with the Institute of British Geographers); Mirumachi, N. (2008) 'Domestic issues in developing international waters in Lesotho: Ensuring water security amidst political instability', in N.I. Pachova, M. Nakayama and L. Jansky, eds, *International Water Security: Domestic Threats and Opportunities*, pp. 35–60, © 2008 by the United Nations University, published by the United Nations University, reproduced with the permission of the United Nations University; Mirumachi, N. (2007) 'The politics of water transfer between South Africa and Lesotho: Bilateral cooperation in the Lesotho Highlands Water Project. *Water International*, vol. 32, no. 4, pp. 558–570, published by Taylor & Francis, reprinted by permission of the publisher (Taylor & Francis Ltd, www.tandfonline.com); and, with kind permission from Springer Science+Business Media: Mirumachi, N. (2012) 'How domestic water policies influence international transboundary water development: A case study of Thailand', in J. Öjendal, S. Hansson and S. Hellberg, eds, *Politics and Development in a Transboundary Watershed – the Case of the Lower Mekong Basin*, pp. 83–100, © Springer Science+Business Media B.V. 2012. I received useful feedback on an earlier version of Chapter 6, presented at the PEAS Research Group meeting at the Department of Geography, during my stay at the National University of Singapore in summer 2013 with the King's–NUS Partnership Award. I also thank Helen Ingram and Larry Swatuk for providing constructive comments on the manuscript.

Writing and (importantly) finishing a book is a long process. Rebecca Farnum and Judy Wall provided editorial assistance at various stages, and Kristofer Chan extended his GIS and cartographic skills in preparing the figures. I acknowledge IWA Publishing and Black Rose Books for permission to reprint Figures 2.1 and 3.1 respectively. One person in particular has seen me enjoy, and struggle with, the process of germinating ideas and putting words on paper. Mathias has always been curious about what I question and research, and why and how I intend to actually make a change. I am ever grateful for his support.

List of abbreviations

ADB	Asian Development Bank
ANC	African National Congress
ASEAN	Association of Southeast Asian Nations
CFA	Cooperative Framework Agreement
cusec	cubic feet per second
DEDP	Thailand Department of Energy Development and Promotion
DFID	Department for International Development, Government of UK
DPR	Detailed Project Report
DWA	Department of Water Affairs, Government of South Africa
DWAF	Department of Water Affairs and Forestry, Government of South Africa
ECAFE	United Nations Economic Commission for Asia and the Far East
EGAT	Electricity Generating Authority of Thailand
EGCO	Electricity Generating Company Limited
EIA	environmental impact assessment
ESCAP	United Nations Economic and Social Commission for Asia and the Pacific
FHH	Framework of Hydro-Hegemony
GAP	*Güneydoğu Anadolu Projesi*; Southeastern Anatolia Project
GBM	Ganges-Brahmaputra-Meghna
GDP	gross domestic product
GEF	global environment facility
GW	gigawatt
ha	hectare
IFI	International Financial Institution
IMC	Interim Mekong Committee
IPP	Independent Power Producer
IR	international relations
IWRM	Integrated Water Resources Management
JCWR	Nepal–India Joint Committee on Water Resources
JPE	Joint Group of Experts
JPTC	Joint Permanent Technical Committee

KCM	Kong-Chi-Mun
km^2	square kilometres
km^3	cubic kilometres
KW	kilowatt
LHDA	Lesotho Highlands Development Authority
LHWC	Lesotho Highlands Water Commission
LHWP	Lesotho Highlands Water Project
m^3	cubic metres
m^3/sec	cubic metres per second
MC	Mekong Committee
MENA	Middle East and North Africa
MOU	Memorandum of Understanding
MRC	Mekong River Commission
MW	megawatt
MWG	Mekong Working Group
n.d.	no date
n.p.	no page
NGO	non-governmental organization
NMC	National Mekong Committee
ORASECOM	Orange–Senqu River Commission
OVTS	Orange Vaal Transfer Scheme
pers. comm.	personal communication
PNPCA	Procedures for Notification, Prior Consultation and Agreement
PoE	Panel of Experts, Lesotho Highlands Water Project
R	Rand (South African currency)
RBO	River Basin Organization
SAARC	South Asian Association for Regional Cooperation
SADC	Southern African Development Community
SAGQ	South Asian Growth Triangle
SAWI	South Asian Water Initiative
SEA	Strategic Environmental Assessment
SPP	Small Power Producer
TCTA	Trans Caledon Tunnel Authority
TFDD	Transboundary Freshwater Dispute Database
TWINS	Transboundary Waters Interaction NexuS
TWP	Thukela Water Project
UN	United Nations
UNDP	United Nations Development Programme
UNEP	United Nations Environment Programme
UNESCO	United Nations Educational, Scientific and Cultural Organization
UNWC	United Nations Convention on the Law of the Non-Navigational Uses of International Watercourses
US$	United States Dollar
WUP	Water Utilization Programme

1   Introduction

How water becomes political

Wicked water

Water resources management has often been described as a 'wicked problem', defying easy solutions. It is wicked because there are unknown dimensions to the science of the natural resource. In addition, there are multiple stakeholders who hold an array of values inherent in water resources management, rendering decision-making difficult (Smith and Porter 2010). International transboundary river basin management presents an even more wicked problem because these rivers are shared by two or more sovereign states, causing decision-making to be all the more complex. Much of the water for human consumption comes from rivers, and there are 276 international transboundary river basins in the world (De Stefano *et al.* 2012; see also Figure 1.1). As major sources of freshwater, many of the world's river basins are already over-exploited, giving rise to concern about water scarcity and diminished economic activity (2030 Water Resources Group 2009). Scientists have warned of the risk of conflict if threats to both the biodiversity of rivers and human livelihoods are not fully understood and addressed with appropriate means (Vörösmarty *et al.* 2010: 560).

To address this wicked problem, there are global calls for water cooperation. The United Nations International Decade for Action 'Water for Life' highlighted water cooperation as one of the key themes between 2005 and 2015. The various policy reports and awareness-raising campaigns of this initiative argue that cooperation brings about efficiency in water resources utilization, spur on regional cooperation and provide broader political bene-fits not just on water issues (see UN Water 2013a). With the increasing concern for impacts of climate change in international transboundary river basins, cooperation is presented as necessary to limit such impacts and to achieve development goals (World Bank 2010b). A state-of-the art report by the UN agencies commented: 'The unavoidable reality that water resources do not respect political boundaries demonstrates the supranational dimensions of water, and represents a compelling case for international cooperation on water management' (WWAP 2012: 32). According to the authors of these reports and protagonists of the global calls, international transboundary river basins would seem to be the epitome of cooperation.

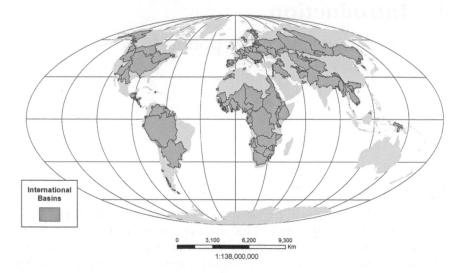

Figure 1.1 Map of international transboundary river basins in the world (Source:
       Product of the Transboundary Freshwater Dispute Database, Department
       of Geosciences, Oregon State University. Additional information about
       the TFDD may be found at: www.transboundarywaters.orst.edu).

However, I question this fixation on cooperation in international transboundary river basins. The imperative for cooperation seems to obscure the
very wickedness and messiness of dealing with these shared waters. Regarding
the Nile River, the cooperation fostered by the Nile Basin Initiative, an institution established between ten basin states, is claimed to be a de facto issue:
'nobody in the basin any longer questions whether cooperation on the Nile is
necessary, desirable or doable. Rather, the conversation has shifted focus onto
how to promote and expedite it' (Seid *et al.* 2013: 37). But what specifically
does cooperation mean for the basin states? What does 'Nile cooperation' look
like? The relationship between Ethiopia constructing the Grand Ethiopian
Renaissance Dam and downstream Egypt poses some interesting questions over
such ideals of cooperation. The dam represents an opportunity for the economic development for Ethiopia, symbolizing its ambition and prowess in the
region (Wuilbercq 2014). For Egypt, the dam raises concerns for its national
security, with the premise that '[i]f Egypt is the Nile's gift, then the Nile is a
gift to Egypt' (BBC 2013). This bilateral relationship has experienced stalled
negotiations over downstream water availability and environmental impacts (Al
Jazeera 2014). The issue flared up at one point in 2013, with the Egyptian
prime minister invoking spilling blood over water (George 2013). Unlike the
claim that cooperation is unquestioned, the political reality presents a complex
situation in which the distinction between whether cooperation will happen
and how it can happen is not so black and white.

While there are calls for cooperation in international transboundary river basins, it is worth examining the contrast between the number of these rivers and levels of institutionalization achieved. The 276 international transboundary river basins are shared by 148 sovereign states, and over 2.7 billion of the world's population is reliant on these waters (De Stefano *et al.* 2012: 198). Acute militarized conflict between basin states is non-existent (Yoffe *et al.* 2003). Rather, between 1820 and 2007, approximately 688 agreements, amendment documents and protocols on international transboundary rivers were signed. Despite this volume of agreements, fewer than half of the international transboundary river basins have established treaties. In basins with three or more sovereign states, there is a tendency for treaties to be signed by only a partial group of states, and treaties that govern the entire basin area in the minority (Giordano *et al.* 2014). Rather than asking how cooperation may be improved, it is first necessary to understand what is happening in a situation where there is no overt conflict but agreements are being signed, though the reach of cooperation is limited by the number of formal treaties. What kind of transboundary water management is being achieved in this situation? What is the relationship between conflict and cooperation?

A critical examination of conflict and cooperation is required if we are to improve our knowledge of international transboundary river basin management. I argue that conflict and cooperation coexist in these basins, and that a narrow focus only on cooperation misses out on a full understanding of the politics of transboundary waters. The focus of this book is the politics that surround the use, management and governance of international transboundary river basins. The main aim is to understand how, and why, conflict and cooperation occur during the process of addressing shared water resources. This process poses serious allocative challenges, not only for the natural resource itself, but also for the values associated with its use and the social order established to manage it. Coexisting conflict and cooperation is a reflection of this socio-political process.

Understanding conflict *and* cooperation over international transboundary rivers

Examining water conflict or water cooperation only would provide a partial picture of how water becomes political. The aforementioned case between Ethiopia and Egypt cannot be simply and sensationally described as one of 'water war' as the media has done (see Schwartzstein 2013). Instead, the project represents how diplomatic conflict through provocative words coexists with deliberations over project details, in an attempt to maintain and challenge water utilization and allocation in the Nile. In the lower Mekong River basin, Laos, Thailand, Cambodia and Vietnam have been celebrated for achieving a water agreement based on sustainable development principles and for establishing the Mekong River Commission (MRC). However, the implementation of the agreement has been controversial, exemplifying degrees of conflict coexisting with the cooperative outlook of the MRC mission. Consensus for basin development plans was lacking among the four states, compounded by an institutional limitation of the MRC and confusion among the development agencies funding it (MRC 2013: 35–36). A critical analyst would point to this mix of conflict and cooperation to explain how riparian relationships are not a given and are subject to a range of factors that shape visions of basin development and 'water cooperation'.

As Wolf (2007) noted, while acute militarized conflict has not occurred between basin states, water scarcity and degradation of water quality can cause political tension between states and instability within states. Even in a comparatively developed region like the Danube River basin, environmental degradation and water quality issues have led to inter-state disputes in the past (Jansky *et al.* 2004). Flooding or water abundance can also be a cause for conflict, especially as many of the existing transboundary water institutions are not sufficiently equipped to deal with flood management (Bakker 2009). Water scarcity is often suggested as a determinant, if not one major factor that makes water allocation and utilization challenging, thus risking conflict. It is argued that competition for limited water availability intensifies political relations between states that may already be fragile (Haftendorn 2000). The Middle East is a frequent example of a region with the high potential for conflict as a result of water scarcity (e.g. Kliot 1993; Lowi 1993; Wolf 1995; Haddadin 2001; Sosland 2007).

Importantly, Mehta (2010c) contended that 'scarcity' may be discursively framed to justify certain solutions, such as technological innovation and fixes. For example, building dams as a viable solution utilizing advances in engineering may serve this discourse of water scarcity. By doing so, scarcity is naturalized, to be taken for granted and as something universal, leading to 'self-fulfilling prophesies around "crisis"' (Mehta 2010a: 252). These expressions of scarcity expose state power manifested with policies and institutions to manage 'crisis', supporting what Michel Foucault called governmentality (Mehta 2010b).[1] The effect of governmentality is that, through discourses of water scarcity, control of society is achieved by the state in a way that is not imposed in an overtly top-down fashion, but through a more pervasive manner (Foucault 1991). However, building more and 'better' dams to deal with water scarcity may in fact only exacerbate the rate of water abstraction beyond ecological limits, leading to loss of livelihoods and other socioeconomic problems. In fact, scarcity is now a hegemonic concept so uncontested that it is seen by many stakeholders as 'almost universally applicable to the water-related challenges of the 21st century' (Sneddon 2013: 17). The observation by Trottier (2003) usefully points out that there are decision-makers who gain from the 'water war' discourse drawing on such notions of scarcity because it consolidates their political power and benefits their agendas. She also emphasized that cooperation expressed as 'water peace' can be manipulated to advance one's interests, allowing it to become a hegemonic concept.

It thus becomes evident that by itself, water is not about politics. As Warner and Wegerich (2010) usefully reminded us, water is not political by default and not always about conflict. Rather, water acts as the medium through which politics occur. They point out that moments of water scarcity, abundance and degraded water quality often give rise to situations of conflict, leading to politics over water. Naturalized, universalized and hegemonized perspectives of scarcity, abundance and other problems relating to water resources are projected and contested by those with a vested interest in international transboundary river basin management. I argue that this politics over water reflects discourses of water as a threat, opportunity, or even a non-issue that allows for both conflict and cooperation to coexist within the relationships between states. Analysis will reveal how 'cooperation' may in fact pose a discursive space where no alternative water management practices or policies are entertained – a position in which decision-makers from Nepal found themselves vis-à-vis India over river basin development plans, explored in Chapter 4 (this volume). It is not necessarily the physical state of the river basin that brings about conflict and cooperation but rather the socio-political process that attempts to determine the solutions and means to address water scarcity, abundance, degradation and environmental stewardship.

A new analytical approach to transboundary water politics

To support the claim that conflict and cooperation coexist in transboundary water politics, I develop an original analytical framework that draws upon several academic disciplines. Scholarly work on the relationship between water and conflict/cooperation has, in large part, been influenced by environmental security studies in the wake of the new post-Cold War world order. The discussion of the management of this natural resource is difficult to tease out from issues of security. For example, a report for the US Department of State expressed the reasons for its concerns about water and security:

> We assess that during the next 10 years, water problems will contribute to instability in states important to US national security interests. Water shortages, poor water quality, and floods by themselves are unlikely to result in state failure. However, water problems – when combined with poverty, social tensions, environmental degradation, ineffectual leadership, and weak political institutions – contribute to social disruptions that can result in state failure.
>
> (Intelligence Community 2012: 3)

The academic disciplines of international relations (IR) and political science have an important legacy in the development of studies on conflict and cooperation over international transboundary river basins. Quantitative work has examined the correlation between conflict and water, and the role of international environmental agreements (e.g. Toset *et al.* 2000; Mitchell and Hensel 2007; Brochmann and Hensel 2009, 2011). Theory of environmental regimes has given rise to qualitative studies of transboundary water institutions and international River Basin Organizations (RBOs). (e.g. Marty 2001; Schmeier 2013).

However, charges have been laid on the limitations to the IR approach that has been the mainstream of international transboundary water analysis. Furlong (2006) argued that using theories of IR results in defaulting to viewing the state as the main actor engaging in conflict or cooperation. She critiqued that hydropolitical analysis is based on a narrow analysis of regimes and hegemony, and that it is detached from and ignorant of the ways in which conflict and cooperation play out at community and individual levels. Instead, other approaches such as political geography and political ecology are much more sensitive to these issues. However, as pointed out by Warner and Zeitoun (2008: 804–805), the existing literature is not only limited to a neo-realist, positivist school of IR that Furlong accentuated in her review, but also includes other perspectives; for example, studies with Gramscian underpinnings of critical international political economy analysis. Perhaps it is not so much that we get rid of IR altogether in hydropolitical analysis. I agree with Warner and Zeitoun (2008) that there is still scope for IR to make useful contributions and particularly from critical schools of thought, particularly on systematic, qualitative power analysis that has been sorely missing in many existing studies. But Furlong's critique does ask for a rethink on how agency is dealt with. An interdisciplinary approach may be more fruitful than simply an IR approach or a political ecology/political geography approach.

This book acknowledges the major contributions made by IR and political science to the shaping of current knowledge about international transboundary river basins; of particular use is the notion of power in politics, as will be explained below; but I will point out the deficiencies of the existing studies that analyse conflict and cooperation separately. Tracing existing IR theories onto empirical contexts is not sufficient because problems of water resources management in shared river basins become simplified through an analysis of conflict *or* cooperation. I present a new approach: the Transboundary Waters Interaction NexuS (TWINS). This approach does not place conflict or cooperation at the centre of analysis. Rather, the analytical focus is on transboundary water interaction: the process through which actors engage over (real and constructed) issues of water scarcity, abundance and watershed degradation. An analysis based on TWINS exemplifies the changing intensities of coexisting conflict and cooperation and seeks to explain how these changes occur. The conceptual framework of TWINS avoids labelling a basin in either conflict or cooperation because transboundary water interactions are not static. They are influenced by a range of factors both within and beyond the watershed (Zeitoun and Mirumachi 2008). The implication is that analysis must also include scalar considerations of what happens at the domestic, regional and basin-wide levels. This new approach aims to address a major gap in the conceptual development of transboundary water analysis.

While IR and political science has been the mainstay of analysis, other disciplines such as new institutional economics, geopolitics and critical geopolitics, economic geography and political ecology have provided important insight for the shaping and evolution of the politics of shared basins (e.g. Bakker 1999; Tvedt 2004; Sneddon and Fox 2006; Dombrowsky 2007; Wong 2010; Myint 2012; see also Agnew 2011). The TWINS framework that I put forward focuses on the spatial scale of water resources use and management. The study draws upon political geography to show ways of refining the understanding of politics of shared waters by looking across spatial scales, not just at the international, inter-state level. Examining the political economy of water resources use at the subnational level also helps explain transboundary water politics. As it will be explained in Chapter 6, the political economy of water in the Mekong River basin is driven by a range of factors, including domestic and regional food and energy demands. In addition, considerations about discourse of water allocation and utilization, and geographical imaginations of the river and the river basin render important explanations on the process of coexisting conflict and cooperation. The development of the Orange–Senqu River basin explored in Chapter 5 is closely associated with geographical imaginations of an untamed, unharnessed and unruly river.

Unpacking the state and state power

Capturing the essence of the political nature of transboundary waters requires a broader range of analytical lenses and a reflexive use of them. This study focuses on transboundary water interactions at the international level as a starting point, in large part shaped by and concurrently shaping the action of states. Rather than assessing events of conflict or cooperation, I examine how actors construct transboundary water interaction. This constructivist approach aids understanding of the way in which state decisions are influenced by knowledge, norms and the worldviews of a particular group of actors: the hydrocracy and politicians that represent state interests. Identifying the ways in which these actors deliberate on water resources avoids analysis of conflict and cooperation that generalizes the state. This is important, especially when in fact it is a group of elites within the state that often shape and gain from 'state interests'. This point addresses Furlong's (2006) critique of the 'state as container'.

By using the concept of the hydrocracy, I further explore the make-up of this elite group and why they matter. The hydrocracy comprises governmental agencies responsible for the use, development and (nowadays) conservation of the water resources. So far as international transboundary water projects are concerned, the organizational reach of the hydrocracy may include agencies dealing with agriculture and food, hydropower and energy, the financing of infrastructure projects and foreign affairs. The hydrocracy is characterized by its preference for engineered solutions for controlling nature based on a belief in scientific and technological progress (Wester 2008: 10). Heads of state, ministers and other political figures not only represent the interests of their nation in international fora and negotiations, but also steer and support the decisions of the hydrocracy. Thus politicians are important in decisions of storing, transferring and distributing water in international transboundary river basins (Molle *et al.* 2009). These elite decision-makers promote the hydraulic mission aiming to capture water for increased irrigation capacity, to prevent disasters caused by flooding and to increase hydropower capacity (ibid.: 333). Extensive and intensive water capture is conducted through the investment of hydraulics and other infrastructure (Allan 2001). Engineering intervention and technology facilitate the construction of dams, canals, wells and desalination plants. These projects are justified because they symbolize the development of the state and foster national prestige. The hydraulic mission requires the mobilization of knowledge, expertise and human resources, in addition to large amounts of investment (Molle *et al.* 2009).

Analysis of the transboundary water interactions through these elite decision-makers provides the opportunity to examine the wider political context in which they operate. This entails consideration of the broader diplomatic context in which negotiations over water resources occur. As the following chapters will show, examining international transboundary water issues as simply a matter of 'low politics', unrelated to foreign affairs and state survival, is not helpful. There are cases where water is securitized during transboundary water interaction, meaning that the usual political processes for decision-making are bypassed and urgent measures put in place (Buzan *et al.* 1998). In other words, water is understood as something existential for the state by these elite decision-makers. Such a situation leads to increased intensities of conflict and reduced intensities of cooperation. This was the case in transboundary water interactions between Nepal and India over a relatively small hydraulic project on one of the Ganges tributaries (explained in Chapter 4, this volume). What are the geopolitical and political factors that trigger such securitization? How does a water issue become so vital to state survival, and why? What do the elite decision-makers stand to gain or lose? Answering these questions requires an understanding of the political context in which shared waters are discussed and a detailed discourse analysis of the hydrocracy and its associated actors.

I argue that these elite decision-makers play a role in shaping basin asymmetries. States within a hydrological basin are not usually equally endowed with regard to water resources, and not necessarily equally capable of deploying adaptive capacity to deal with water scarcity and water abundance (Kistin 2010). The focus on elite decision-makers is one way of examining how such differences are dealt with in the pursuit of specific interests and motivations to use the shared waters. These elite decision-makers draw in other actors such as development banks, and private businesses in construction and consultancy (Molle 2008: 218). This hub of actors, satelliting around the hydrocracy and politicians, begins to shape national interests. In this way, the study places power as a key analytical component to understanding transboundary water interactions. The Framework of Hydro-Hegemony (FHH) developed by Zeitoun and Warner (2006) is used to explain asymmetric relationships between basin states. Basin states that are hydro-hegemons maintain asymmetry through the relative advantage of three factors: riparian position, hard and soft power, and the capacity to physically capture waters (Zeitoun and Warner 2006: 443–452). I further refine the FHH by providing more specificity on agency. While the FHH does not differentiate actors that maintain and challenge basin asymmetry, I use the notion of material capability and discursive power to exemplify how these elite decision-makers operate. In this study, material capabilities are understood, first, as a representation of riparian position, and second, as the elite decision-makers' capacity to capture waters that assist hydraulic control of shared water bodies – whether it be financial resources or technical expertise. Power asymmetry is explored in the form of the discursive power of these decision-makers. Such discourses, in addition to the water management practices spearheaded by the hydrocracy, reveal the asymmetric control over water resources that exists between basin states.

Taken together, material capacity and discursive power exemplify how hydro-hegemony is represented in transboundary water interactions. The focus on the hydrocracy, politicians and associated elite decision-makers helps unpack the exercise of state power. This is not to say that non-state actors do not matter or are exempt from a role in decision-making. Rather, the intent with the focus on elite decision-makers is to present the deeply entrenched milieu in which these actors set, implement and monitor national policies and agendas relating to water resources management and governance. These policies and agendas are driven by the pressures on water supply and demand, and by sustainable development notions. It is important to consider the influences of such national considerations on and from basin-wide multilateral negotiations. The hydrology of the transboundary river, which ignores the artificial political boundaries, binds basin states to a very awkward unit of management and governance. However, it is not the material hydrological bounds that determine the management and governance but the wider political context in which transboundary water interactions play out, and the evolving political economy of water resources use. These political and political economic contexts give rise to geographical imaginations of the river and river basin promoted by the elite decision-makers.

Understanding the role of power and hydro-hegemony is important because it can help explain how cooperation – in the form of an international agreement or a joint technical committee – can allow uncertainty over water allocation to exist and persist. Superficial cooperative initiatives may be used, not to fundamentally resolve the issue but rather to put any change to the situation on hold. We should question how cooperation is justified, thereby looking into their 'depth' at dealing with such uncertainty: What does 'shallow' or 'deep' cooperation do? By using the concept of hydrocracy and focusing attention on agency, this study furthers the use of power analysis and presents a more detailed analysis on who uses what kind of power, how and why. Analysis of power relations provides persuasive explanations as to how hydrocracies and politicians establish and impose decisions over water allocation and use, and why those decisions may not be equitable for all, especially for those relying directly on the water for their livelihoods. Therefore, an analysis of transboundary water politics cannot explain these nuances if it only focuses on events that represent conflict or cooperation, and not on transboundary water interactions (Zeitoun and Mirumachi 2008). The empirical examples in this book will demonstrate how conflict and cooperation coexist in this process, how they change in their intensity, and why power, particularly discursive power, is crucial.

Shared blue waters in the Global South

The analysis presented in this book is particularly concerned with transboundary water politics in the developing world. Gupta and Lebel (2010) argued that water resources management needs to contend with the dual challenges of access and allocation: access to drinking water and sanitation; and allocation of water for multiple economic, social and ecosystem purposes, and of responsibilities and risks of such water uses. The inclusion of water issues in the Millennium Development Goals was a reminder of the importance this natural resource has for economic development. None the less, the poor continue to be disproportionately affected by challenges to water access and allocation (UNDP 2006; WWAP 2009; Vörösmarty et al. 2010). The case studies in this book on the Ganges and Mekong river basins deal with the issues of allocation. While the analysis of this book does not deal with the issue of individual access to water, it does engage with river basin development projects to secure drinking water, as in the case of the Orange–Senqu River basin. The analytical approach to coexisting conflict and cooperation is used to draw out issues relating to the way in which water is stored, transferred and distributed in shared river basins.

Abstraction from and the appropriation of international transboundary rivers for human activity transforms the landscape and the geography of a basin. In addition, this process prompts further socio-economic changes to occur with a particularly large impact on the environment and local livelihoods in developing countries. Blue water (that is, water from rivers, lakes and aquifers) is very valuable for agriculture. More than 40 per cent of food production relies on irrigated blue water (WWAP 2012: 46). Blue water is used for agriculture less than green water, or soil moisture from precipitation for agriculture, but the storage of the latter is difficult. Rivers provide opportunities for storing blue water to deal with variability. Such storage also enables energy production with hydropower dams. Compared to green water, managing blue water requires consideration of providing access to and allocation of multiple purposes. Governing blue water is thus closely associated with policy decisions on agriculture, land use, environmental stewardship, energy, and human development. Managing international transboundary rivers therefore has implications for food, energy and human security.

A study published in 2003 identified 17 international transboundary river basins that were in danger of facing political contention within the next decade, many of which are in the Global South (Wolf *et al.* 2003). The Ganges–Brahmaputra, Han, Kura–Araks, Ob, Salween and Tumen rivers in Asia, and La Plata and Lempa in Central and South America, were mentioned. In Africa, the Incomati, Kunene, Lake Chad, Limpopo, Okavango, Orange–Senqu, Senegal and Zambezi rivers were highlighted. More than a decade later, there has been no acute militarized conflict between the basin states. In fact, a follow-up study showed that many of the general trends of conflict and cooperation in basins around the world have stayed the same (De Stefano *et al.* 2010). Rather, the implication of these examples is that the risk of (non-militarized) conflict is high in developing countries because states lack robust institutions to manage these waters (Wolf *et al.* 2003).

However, I argue that these basins are facing challenges that cannot merely be captured by the effectiveness of institutions. For example, there is an increasing interest in the Salween River as new markets for energy and trade open up with Myanmar moving towards democracy. Similarly, hydropower expansion is being sought on the Uruguay River in the La Plata River basin. In the Orange–Senqu River basin, new infrastructure for bilateral water transfer is being constructed. Understanding how elite decision-making brings about the hydraulic mission that controls river flows is central to the analysis I present in this book. At the same time, the achievement of the hydraulic mission is subject to wide-ranging drivers for water access and allocation.

Looking beyond the boundaries of IR becomes important because problematizing the governance, management and the day-to-day practice of international transboundary river basins is elusive. Different stakeholders will have different ideas concerning the challenges the basin is facing. Flooding may be interpreted as a technical issue requiring better forecasting and warning systems by governmental actors, while it may be seen as a livelihood issue by local communities. In transboundary river basins in the Global South, the complexity of these perspectives on river basin development is further magnified by the involvement of donor agencies, International Financial Institutions (IFIs), private sector investors, and project developers and businesses. Controlling the river flows by elite decision-making to achieve the hydraulic mission brings about questions relating to allocation and utilization: how is decision-making on the allocation and use of shared waters justified? What are the implications of such decisions for the individual basin states and for the basin as a whole?

Ganges, Orange–Senqu and Mekong river basins

Through case studies of the Ganges, Orange–Senqu and Mekong river basins, this book examines the ways in which developing countries deliberate, negotiate and implement river basin management. These three river basins were selected for analysis because they enable the research questions to be answered insofar as they have similar socio-economic development contexts. All three basins are made up of states that are in various stages of economic diversification and development. In other words, the basins are facing challenges of hydraulic development and going through changes in their political economies of water, and in their political economies in general. Such changes provide opportunities for conflictual and cooperative interactions driven by political, socio-economic and institutional factors.

Part of the Ganges–Brahmaputra–Meghna River basin, the Ganges River is shared by Tibet, Nepal, India and Bangladesh. A main feature of this river basin is the highly variable flow during the year. The distinct wet and dry season flow has brought about a unique set of water allocation and utilization problems. In particular, between Nepal and India, key issues have been on hydraulic infrastructure development to take advantage of water abundance and on the prevention of hazards (Dixit 1997).

The Orange–Senqu River basin is located in Southern Africa and is shared by Lesotho, South Africa, Botswana and Namibia. The Vaal and Senqu Rivers are the main tributaries, enabling South Africa and Lesotho to each control the headwaters with these separate tributaries. The geographical location of South Africa makes it a unique riparian: it is a downstream state vis-à-vis Lesotho, but an upstream state vis-à-vis Namibia. Because the basin spreads over a number of climatic zones, there are large temporal variances of water flow within the basin area of South Africa. Securing 'food water' (Allan 2013) or water for agriculture is an issue in these conditions. However, access to 'non-food water' (industrial and potable water (Allan 2013)) is also a major concern, particularly for South Africa. The fact that urban centres are located away from the main water sources makes water access an important issue, especially for the Gauteng province where large cities such as Johannesburg and Pretoria are located (Turton *et al.* 2006).

The Mekong River basin in Southeast Asia is described as being at a 'crossroads' (Keskinen *et al.* 2012). The beginnings of a major hydropower boom has generated debate on the ways in which the river should be developed. However, the basin has long been the setting for multilateral, bilateral and national projects for navigation, irrigation, hydropower, fisheries and flood management. This basin is also unique in that between the lower basin states, Laos, Thailand, Vietnam and Cambodia, there is a history of institutional development addressing water allocation and utilization. While the upstream states, China and Myanmar, have hitherto not participated in the river basin institutions, the basin-wide implications of these development plans can be far-reaching, well beyond the existing institutional range.

These three cases represent both wet basins (Mekong, Ganges) and semi-arid basins (Orange–Senqu) in which water allocation and utilization impact upon transboundary water interactions. In addition, the case studies enable a consideration of hydro-hegemonic states that are both upstream and down-stream in particular bilateral relationships. For example, between Thailand and Vietnam, the former is the upstream hydro-hegemon. Between Lesotho and South Africa, the latter is the downstream hegemon, as is the case between Nepal and India. The three basins are treated as explanatory case studies that probe causal relationships, and help test analytical approaches (Yin 2009), in this case of TWINS.

Transboundary water politics: understanding environment and society

At the heart of transboundary water politics lies the question of equity. As the empirical analysis of the Ganges, Orange–Senqu and Mekong river basins contained in this book demonstrates, water allocation between states is a major issue. Equitable water allocation in international transboundary river basins is particularly contentious, with basin states claiming their sovereignty and entitlement to utilize the waters in their territory. Water demands often draw upon the development needs of the state. These claims are supported or subverted by basin asymmetries, resulting in a situation where water resources management brings about an outcome that is 'equal' so far as the hydro-hegemonic state is concerned (Zeitoun *et al.* 2008, 2011).

This issue of equity is common to many basins around the world, as was also seen in the recent events in the Nile River basin explained earlier. Lengthy processes of attempts at water governance have been made but are shaped by unique circumstances. A 'historic acquired right' has been claimed by downstream Egypt to utilize the waters, and a bilateral treaty of 1959 between Egypt and the Sudan assumed that an overwhelming share of water resources of the basin was the entitlement of these two states, at the expense of the other upstream states (Cascão 2008). This claim to entitlement has been a stumbling-block to establishing legal and institutional principles for water cooperation through the Cooperative Framework Agreement (CFA) since the late 1990s. While upstream states aimed to revise historical alloca-tion agreements, the CFA details have been criticized as woefully flawed, only to maintain existing asymmetrical water allocations (Mekonnen 2010). The controversy over the Grand Ethiopian Renaissance Dam is set against this intractable issue of allocation and equity of long duration. However, with movements spurred by the Arab Spring in 2010, the political landscape of the Nile River basin is changing, not to mention the existence of a new basin state, South Sudan. New leaders, new political regimes and new ambitions over the Nile waters are evident. How material and discursive power is used to legitimize this new dam will mark another phase of transboundary water interactions in the Nile River basin.

There are various mechanisms that guide equitable water allocation and utilization: quantitative allocation; percentage allocation; benefit sharing. To address the increasing concern about the global water crisis, international legal frameworks like the United Nations Convention on the Law of the Non-Navigational Uses of International Watercourses (UNWC) have been established to foster cooperation over shared waters. Adopted by the UN Assembly in 1997, the UNWC has laid out several key principles regarding water allocation and utilization, including the principle of equitable and reasonable use. Through this principle, the UNWC recognizes equal rights of basin states, whatever their geographical position, to utilize shared waters. In addition, this principle requires both biophysical and socio-economic considerations for the assessment of needs for water use (Rieu-Clarke *et al.* 2012: 104). The UNWC is a major achievement in establishing a global norm, and was ratified by 35 states, finally entering into force in 2014. However, its implementation cannot be divorced from the messy, political reality in which legal principles are referred and ignored by basin states through the exercise of different forms of power (Mirumachi *et al.* 2013).

The study of transboundary water politics and the underlying issues of inequality have a strong policy dimension. Jimenez and Perez-Foguet (2009: 4) found that despite promises of increased official development assistance funds, in practice assistance to the water sector decreased between 1995 and 2004: the accumulated value was US$46.3 billion. In addition, they found that the distribution of this assistance was skewed in comparison to water demand. For example, South and Central Asia and Sub-Saharan Africa received only a combined 32 per cent of total investment, despite these regions accounting for 61 per cent of all people without access to water (ibid.: 7). And a study by Nicol *et al.* (2001) found that even though institutions were being established to manage international transboundary river basins, funding for their establishment remained limited. Sandler (2006) identified similar difficulties when he found that international donors channel funds into national public goods rather than into regional public goods, such as international river basin organizations. If there is insufficient funding for transboundary water management, it is all the more important that institutions and policies are guided by scientific insight regarding the nature of transboundary water interaction.

Furthermore, transboundary water politics reveals issues of inequalities at local and national scales as a result of inter-state interactions. Dam projects, irrigation expansion, inter-basin transfers and the pursuit of the hydraulic mission expose serious trade-offs between project-specific benefits and other modes of economic development. Much of the immediate and long-term impact of such projects is felt most acutely at the local level. Costs and benefits are discussed by decision-makers, often far removed from the geographical locations of these projects. Project calculations and assessments can reduce the local geographies to a homogenized, general socio-economic and environmental context. This in turn brings about questions of environmental and social justice; of who benefits from the environmental goods and who is affected by the environmental bads. The analysis of transboundary water politics is a useful way to begin questioning the accountability mechanisms to address these trade-offs, the appropriate scale at which these mechanisms should exist and the scalar implications.

Often, the treaties, RBOs and other institutions established in international transboundary river basins refer to sustainable development. The use of sustainable development as a guiding principle considers the balance between economic development from using the river and the maintenance and conservation of the watershed. However, sustainable development is a concept that is constantly debated, with no concrete definition. Moreover, sustainable development is inherently political and involves consideration of governance systems that embed sustainable development goals and facilitate action (Jordan 2008). The issues concerning international transboundary river basins in developing economies, in particular, pose fundamental questions about development and poverty reduction, and the ways in which society views nature.

Solving the wicked problem of water is becoming an ever-pressing issue as global concerns about the sustainability of water resources mount. These concerns are expressed through discussions of the 'planetary boundary' of freshwater being endangered. As food demand – and thus agricultural water use – increases, human consumption of water resources may cause irreversible harm to the earth system, destroying both ecosystems and human livelihoods (Rockström *et al.* 2009). The hydrological cycle is increasingly influenced by climate variability and change, as well as human intervention through extraction and storage of water and land-use change (WWAP 2012: 80). Dams providing water for agriculture and energy have altered the flow of many large rivers in the world (Nilsson *et al.* 2005). In broad terms, examining transboundary water politics is one way to reflect on society's relationship with nature.

Structure of the book

The arguments outlined above are spread over the next six chapters and make up the body of this book. Why a new approach to the analysis of conflict and cooperation in transboundary water politics is necessary and crucial is explored in further detail in Chapter 2. This chapter takes stock of literature that has advanced our understanding of the politics of international transboundary rivers. The review refutes the persistent claims about imminent 'water wars'. Largely drawing upon IR and political science, the chapter explores the various conceptualizations of conflict and cooperation. The chapter then critiques these conceptualizations, exposing the shortcomings of analysing conflict or cooperation separately.

To address these weaknesses, Chapter 3 develops a conceptual framework that enables critical examination of conflict and cooperation among basin states. The TWINS framework and a conflict–cooperation matrix are introduced.

The theoretical foundations of the TWINS framework and the key concepts to operationalize analysis are explained in detail. Basin asymmetry and power are underscored as important factors to explain both the process and outcome of transboundary water interaction. These concepts are then taken up in the development of a conflict and cooperation intensity scale integral to the TWINS matrix. In particular, securitization and socialization provide theoretical insights to develop a nuanced scale of conflict and cooperation intensities. The chapter also points out the benefit of drawing on the geographical literature, emphasizing the richness that political geography and geopolitics can bring to the analysis through spatial considerations to the factors that shape transboundary water politics.

Having argued why a new approach to transboundary water politics analysis is needed and how that may be achieved, Chapters 4, 5 and 6 apply the TWINS framework to the Ganges, Orange–Senqu and Mekong river basins. Collectively, these chapters show how asymmetries in material capability and discursive power of particular basin states are manifest in the changes in conflict and cooperation intensity as the transboundary water relations evolve. These chapters focus initially on bilateral relationships that best exemplify some of the key challenges the basins face, then move on to consider the basin-wide implications of the bilateral transboundary water interactions.

In Chapter 4, the case study of the Ganges River basin, focusing on Nepal and India, provides an insight into a type of bilateral relationship that is project-based. Through projects to develop the river basin, Nepal and India have established bilateral water relations that have not changed dramatically in conflict and cooperation intensity. However, careful analysis shows that cooperative mechanisms established between the two states actually make it difficult to change the status quo of water allocations. These insights are then used to review multilateral basin governance, especially as water and climate change concerns mount across the Ganges–Brahmaputra–Meghna River system.

The case study of the Orange–Senqu River basin in Chapter 5 focuses on the unique international water transfer project between Lesotho and South Africa. The TWINS analysis shows that geopolitics is important in explaining the nature of the bilateral relationship over water resources. The examination of the domestic politics of water also provides some very persuasive insights into the interests and motivations that inform water allocation and use by elite decision-makers. In contrast to the Ganges River basin, the Orange–Senqu River basin has a multilateral governance scheme in place that integrates all basin states under shared principles of water governance. The chapter conducts a critical examination as to whether such a multilateral scheme can challenge hydro-hegemony and address issues of inequality over water allocation.

In Chapter 6, the Mekong River basin is analysed. Out of the three basins, it may be said that efforts at institutionalizing the Mekong River basin management have been most evident, with RBOs in place since the 1950s. Through the transboundary water interactions between Thailand and

Vietnam, it becomes clear how and why certain norms and principles that guide the RBOs were put in place. This process has been highly political and has constantly exposed the complexity of 'national interests' that shape transboundary water interactions. Transposing this insight to the recent increase in hydropower projects in the basin, the chapter delves into the political economy of dam building, which defies the very structured, hierarchical decision-making nature of the current RBO. Issues of mainstream and tributary dams in the Mekong engage a wide spectrum of state, business and civil society. This case study is useful in showing how hydro-hegemony operates in issues of hydropower, and how decisions in the energy sector impact upon water use.

Chapter 7 returns to the questions set out at the beginning of this book, and explores the wider implications of the findings from the three case studies for further analytical development and policy. First, the chapter discusses the factors that contribute to changing conflict and cooperation intensities. In many cases, while there is no overt conflict over water, high levels of cooperation are also hard to find. A critical explanation to 'shallow cooperation' is provided, demonstrating the mixture of material capability and discursive power. Second, by considering the ways in which power is exercised, FHH is refined. This draws upon insights from the ways in which geographical imaginations and material capability are used. Third, I go into depth about the nature of decision-making by elite decision-makers, the problem frames they construct and the solutions they propose. Such an analysis shows the resilience of the hydrocracy and how it has implications for basin-wide water governance. The chapter offers some policy-oriented thinking on refining transboundary water governance, and concludes by returning to the importance of recognizing the very political nature of transboundary water politics.

Note

1 See also discussions on how water resources management, particularly in the supply of drinking water, is about biopolitics of the government to seek control over citizens in urban areas (Kooy and Bakker 2008; Bakker 2013; Ekers and Loftus 2008).

2 Explaining transboundary water conflict and cooperation

Introduction

Hydropolitics, or the body of work on transboundary water conflict and cooperation (Elhance 1999), provides various explanations for conflict and cooperation from a range of disciplinary perspectives. However, a grand theory of international conflict and cooperation over shared waters has yet to be developed. Theoretical development has been patchy, with Frey and Naff (1985: 70) cautioning in the 1980s against the analysis of conflict and cooperation over shared waters without theoretical guidance. Almost two decades later, Bernauer (2002: 2) critiqued the exiting literature as being 'almost entirely descriptive' and noted that empirical findings did not contribute much to generalizations, despite the growing body of case studies.

Studies on international transboundary river basins provide a fragmented picture on the causal links between water and conflict or cooperation, based on various disciplinary assumptions. In order to refine analysis on transboundary water politics, it is important to acknowledge how different disciplines have contributed to the understanding of inter-state conflict and cooperation. However, as this chapter will show, there are some major limitations in conceptualizing conflict and cooperation. Addressing these deficits in the existing literature is crucial if we are to further our understanding of politics over the allocation and use of transboundary river basins. To this end, I argue that rather than being bounded by disciplinary silos, incorporating plural perspectives on agency, scale and power is necessary for more critical and in-depth analysis.

To take stock, a broad sweep of studies that have provided explanations for inter-state conflict and cooperation over water is presented in the following pages. Explaining transboundary water conflict is perhaps easier than explaining cooperation. Frey (1993) attempted to theorize violent conflict over shared water resources by considering scarcity, power differentials between states and riparian position. The analytical focus of many existing studies may be classified according to these three factors. However, cooperation seems much harder to explain, much less define. As an indication, within IR there are a large number of analytical frameworks of cooperation, with their own definitions, conceptualizations and terminology

(Oye 1985). IR analysis of cooperation has been examined from a set of six hypotheses focusing on the nature, process and actors (Milner 1992). This means that hydropolitical studies drawing on IR has a broad range of dependent and independent variables. The review is not intended to be comprehensive, but rather to show the breadth of explanations that exists in scholarly literature, and to provide the basis of the argument that demands a more critical approach towards conflict and cooperation.

Water scarcity and the importance of water

The quantitative aspect of water supply and demand has been a recurring theme within hydropolitics in understanding conflict over natural resources. Despite it being a renewable resource, water is characterized by its uneven distribution across space and time. Consequently, securing enough water in a reliable fashion is a key challenge for water resource management. Falkenmark (1989, 1990), a hydrologist by training, was one of the first scientists to develop a body of work on water scarcity in an attempt to inform policy, especially with regard to challenges that could be faced in African countries. The premise to her studies was that population increase makes meeting water demand more difficult. Political scientist Homer-Dixon (1991, 1994) modelled the links between different types of natural resources and conflict. He argued that water, as a renewable resource, was prone to militarized conflict triggered by scarcity (Homer-Dixon 1991). His subsequent research, however, showed that scarcity is 'mainly an *indirect* cause of violence, and this violence is mainly *internal* to countries' (Homer-Dixon 1999: 18; emphasis in original). None the less, it has also been argued that scarce water resources can play a role in strengthening the economic or political power of a state, making water scarcity a significant issue. Thus, in some circumstances, water can be the tool for advancing militarized action, or a vehicle through which military or political goals are achieved (Gleick 1993).

However, using absolute water quantity as a factor to explain conflict is problematic. Rather, it is the perceived importance of water that is worth analysing. This is because it reflects the 'value priority or motivation' of the basin state in its interactions with other basin states, as Frey (1993: 61) pointed out. He gave the example of how Turkey came to value the waters of the Tigris and Euphrates Rivers once the Southeastern Anatolia Project (GAP), a large-scale project consisting of multiple dam construction, became an important item on the national agenda. Social adaptive capacity is another concept to explain the social and political process in which water resources are valued and managed by society, thereby placing emphasis not on the physical quantity of water but on institutional determinants (Ohlsson 1999, 2000). Dinar (2009) posited that the level of water scarcity is useful to explain cooperation rather than conflict. In fact, he argued that different levels of scarcity stimulate negotiations between states that explore bargaining strategies for a mutually desirable outcome. In other words, basin states facing moderate water scarcity have the best chance of being able to negotiate and to cooperate; severe scarcity or abundance yields situations that result in low cooperation.

The causal link between water scarcity and conflict must be treated with caution, as it has deterministic assumptions (Lipschutz 1997). A positive relationship between the two is an over-simplified assumption that obscures the inter-state political process of transboundary river basin management. Studies that connect population growth, scarcity and conflict are Malthusian in character. Put differently, Homer-Dixon's work (mentioned above) focuses solely on population growth to explain scarcity, which can be critiqued as providing only a partial picture to the character of conflict over natural resources (Hartmann 2002). Barnett (2000) provided a very persuasive argument about the role played by politics in conflict over water:

> [I]f there is conflict over water, then that conflict is the result of a *failure of politics* to negotiate a settlement over the shared use of water. The idea that a war over water, or any other resource, is not a war about politics is dubious. Politicians and military leaders might wish to present war in Darwinian or Malthusian terms as a fight over subsistence needs, but this 'state of nature' rhetoric is a pragmatic device that denies responsibility for peaceful action, and justifies violence in lieu of meaningful dialogue.
>
> (Barnett 2000: 276; emphasis in original)

Unpacking this 'failure of politics' may be done in several ways, and the political ecology perspective on renewable resources (not limited to water) by Le Billon (2001) is a useful start. His analysis on the scarcity of renewable resources and conflict provides a richer account than the Malthusian or neo-Malthusian one. He argued that the political economy that shapes the way in which resources are extracted and traded, the spatial geography that facilitates control over resources, and the degree of dependence on them, are all parts of the story that explains acute conflict (ibid.: 566).

On the point about political economy, Allan (2001) pointed out that in the arid Middle East and North Africa (MENA) region, states import crops and other food products without having to secure additional water resources to ensure economic development. This provides the basis of why inter-state water wars in the region have not occurred, despite being poorly endowed with water resources. The concept of virtual water exemplifies the political economy in which water embedded in food and other manufactured products flows across political borders (see also Allan 2011). The political economy approach necessitates the analyst to examine the level of economic diversification of the state and the means it has to secure water, both within the basin through engineering solutions and outside the basin through virtual water 'importation' (Allan 2001). In this case, it is not so much the success of politics in ensuring regional cooperation over water resources, but rather a 'politically silent' means that averts military conflict (Allan 2002: 260).

Riparian position and externalities

Geographic proximity to the water source is another factor that has been considered to explain upstream-downstream dynamics in the politics of water allocation and river basin management. Frey (1993: 61–62) contended that whether a state is located upstream, midstream or downstream can contribute to it engaging in conflict or in forming coalitions to secure water. He argued that upstream riparian states have the advantage over downstream states with regard to water access and capture, while midstream and downstream states may form coalitions and influence the upstream state. LeMarquand (1977) also noted the upstream advantage with respect to water allocation. He claimed:

> A state that takes advantage of its favored position on a river has no real economic incentive to alter its behavior. Consumptive use of the river's waters, flow regulation, and waste disposal by an upstream riparian are examples of water use that lead to upstream–downstream conflicts.
>
> (LeMarquand 1977: 10)

Allouche (2005) offered a view on the relationship between water resources, state-building and nation-making that explains why geography is relevant to transboundary water politics. He argued that water resources are closely connected to concepts of national territoriality and sovereignty – to such an extent that states are prepared to engage in conflict over shared waters.

Geographic position brings about issues of externality. While there are valid arguments that question whether the natural boundary of hydrological basins best facilitates the management of shared waters (Warner *et al.* 2008; Venot *et al.* 2011), the norm is to manage the transboundary river basin as a hydrological unit divided by sovereign borders. This is not unusual when managing natural resources: for example, the management of fisheries and forestry is often premised on the boundaries of nation-states. Because of this artificial division, when a state develops a dam, diverts or pollutes the river, this may impact upon the usage of other states. This trans-border effect is described as an externality. Empirical case studies show that externality problems tend to be more challenging than collective problems, because different basin states have incompatible interests and require different incentives (Marty 2001: 346–352).

The concept of externality may be used to classify the kinds of problems and opportunities in transboundary river basins that lead to conflict and cooperation. Using an institutional economics approach, Dombrowsky (2007: 48–52) provided a useful typology of externalities that exist in the management of transboundary waters. In broad terms, she described water abstraction activities as having negative externalities, while hydraulic infrastructure construction has both positive and negative externalities. In addition, these externalities may either be unidirectional in the case of rivers crossing a boundary, or reciprocal in the case of a river forming a boundary. According to her, water abstraction and land-use change in one state usually have a negative effect on the water usage of other states because their ecology is affected. On the other hand, hydraulic development for the control of water flow can provide positive externalities, though hydraulic infrastructure can also have a negative impact upon the ecology, not necessarily benefiting all users in the basin. The typology suggests that riparian states are less likely to engage in cooperation if the problem of externalities is one of negative unidirectional externalities (ibid.: 274). This is because, in that situation, there would be no shared costs or an opportunity to reduce them compared to the problems brought about by reciprocal externalities. It follows that negative unidirectional externalities require allocative arrangements of property rights in order to encourage cooperation (ibid.: 266).

Power

Frey (1993) hypothesized that superior military power can secure and safeguard a state's access to water. In a study of shared waters in the MENA region, Shapland (1997: 164) commented that while war over water has been rare, military strength is, none the less, an important factor in deterring weaker parties from exploiting the river basin. In attempts to secure water access there has been, to date, little use of military force in international river basins (Wolf *et al.* 2003); but analysts have identified other forms of power that has been exerted.

Most notably, Zeitoun and Warner (2006) developed the Framework of Hydro-Hegemony (FHH), in order to explore the way in which power is employed during the negotiation and implementation of water allocation arrangements. They argued that in order to control water resources, states with more power – or the hydro-hegemons – can coerce and induce compliance from a weaker party, through a variety of means. The hydro-hegemons can (1) physically capture the resource, (2) maintain the asymmetry in the basin through treaties and other methods that secure preferential water control, and (3) establish regimes so that if the weaker party does not participate, it would be marginalized (see Strategies in Figure 2.1). Various tactics support these three strategies for resource capture: containment or integration, such as military might; provision of incentives; and limiting alternative discourses (see Tactics in Figure 2.1) (Zeitoun and Warner 2006: 444–450). The main contribution made by FHH is its demonstration of how power determines the outcome of decision-making among basin states in ways that are not necessarily easy to measure or quantify. Importantly, FHH highlights that 'the absence of war does not mean the absence of conflict' (ibid.: 437).

Reviewing existing studies along the lines of water scarcity and its importance, riparian position and associated externalities, power is useful in showing the breadth of explanations given to why acute conflict does not occur but, at the same time, makes cooperation difficult. However, these studies tend to give a piecemeal insight into when and how water becomes political. An

Roman numerals relate to Lustick's (2002) classification of increasingly efficient compliance-producing mechanisms: (I) coercion, (II) utilitarian exchange, (III) instigating normative agreement and (IV) inducing ideologically hegemonic beliefs. The list is non-exhaustive.

Figure 2.1 Strategies and tactics of control over transboundary water resources according to the Framework of Hydro-Hegemony (Source: reproduced from Zeitoun, M. and Warner, J. (2006) 'Hydro-hegemony: A framework for analysis of trans-boundary water conflicts. *Water Policy*, vol. 8, no. 5, pp. 435–460, with permission from the copyright holders, IWA Publishing).

analytical framework that includes problem structures, power, basin-specific and external influences is lacking.

Incentivizing cooperation

Studies that explain the cause of conflict and non–cooperation provide some hints on how cooperation may be induced. First, the perceived importance of water guides studies that focus on its economic value. Whittington *et al.* (2005) proposed that if the economic value of water within a basin were calculated, cooperation would be easier to foster. Assessing the economic value would enable holistic river planning rather than fragmented state-oriented planning, and thereby reducing concerns over the territorial and sovereignty issues of shared waters. By taking a basin-wide economic perspective, their study found that all basin states in the Nile River basin could gain economically through various hydraulic investments (ibid.: 249–251). An analysis of economic values also helps identify economic incentives that can influence the attitude of individual basin states. In the case of the Nile, providing economic incentives for Sudan can outweigh the economic value of unilateral action and thus this midstream state could be the catalyst for basin-wide cooperation (Wu and Whittington 2006: 8–9).

Analysis of international institutions or regimes has enabled exploration of the conditions necessary for cooperation. Studies point out that institutions can be useful in addressing the problem of externalities. Jägerskog (2003) examined the hydropolitical relationship between Jordan, Israel and Palestine, in the Jordan River basin, and concluded that these states had established a water regime because they acknowledged their hydrological interdependence and, thus, the need to manage externalities arising from such interdependence. His study showed that while power asymmetry influences agreement implementation, establishing a water regime fosters cooperation and deters defection. Kibaroğlu (2002) claimed that regime-building in the Euphrates–Tigris River basin may be instrumental to achieving basin-wide agreements. She advocated that with a basin-wide regime, proportionate water allocation can be realized between the basin states, data and information exchange encouraged, and water management practices monitored.

However, other empirical studies reveal patchy results with regard to the influence institutions may wield to induce and sustain cooperation. Kistin *et al.* (2009: 10) argued that while many international transboundary water agreements are in place in Southern Africa, their implementation poses difficulties, such as those related to data transparency. Some explanations may be sought in the way institutions are set up. Schmeier (2013) developed a framework based on institutionalist theory, with the goal of understanding the role of River Basin Organizations (RBO). Her study found that the effectiveness of RBOs depended on their institutional design mechanisms; in other words, it matters how the organization was set up and what its remit includes. Specifically, she argued that its membership, its financial structure and its mandated scope are particularly crucial factors.[1] Utilizing a qualitative comparison of three case studies from the Mekong, Danube and Senegal river basins, Schmeier's study identified that key design mechanisms need to complement each other in order to strengthen the effectiveness of the organizations to solve problems. For example, she demonstrated that in the Senegal River basin, the Oranisation pour la Mise en Valeur du Fleuve Sénégal has not been able to improve water resource management – despite encouraging cooperation between basin states. She identified a major flaw in the institutional design of this RBO: a broad organizational mandate well beyond the scope of water resource management. The particular set-up effectively demands economic development at the cost of environmental degradation, which is at odds with efforts to improve water governance (ibid.: 245–246).

The insight into institutional design is useful, given the global trend to establish such organizations in shared basins, and the growing reliance on their mandate to foster inter-state cooperation. Frameworks of institutional effectiveness may assume power asymmetry (e.g. Berardo and Gerlak 2012), but it needs to be explicitly analysed to understand how determining power can be. It should be underscored that getting the design right *is* political, not just a technocratic exercise. To this end, understanding the power asymmetries helps explain the extent to which institutions can be effective. To use an example to emphasize this point, Waterbury (2002) scrutinized the prospects of, and challenges to, the Nile River basin states. He argued that voluntary cooperation through water management institutions is difficult, even though incentives could be offered by external organizations, such as the World Bank. This means that benefits of cooperation through multilateral institutions are high for some and low for others, making long-term cooperation challenging (ibid.: 173).

In addition, analysts have given much thought to the negotiation process, designed to bring about cooperative outcomes. Negotiation strategies, such as issue linkage, enlarge the scope for inter-state agreement and give rise to 'induced cooperation' (Weinthal 2002: 38–41).[2] It follows that if states can find ways to strategically link water and trade, for example, the prospects for cooperation increase. In the Euphrates–Tigris River basin, Daoudy (2009) found that downstream Syria linked water with security issues during negotiations with upstream Turkey in the 1980s and 1990s. Turkey had planned to develop storage and hydropower dams for the GAP, which would have had a negative impact on water availability downstream. In an attempt to negotiate preferable water allocation, Syria made moves on an issue not related to water, namely the support of Kurdish insurgency movements. Daoudy's case study demonstrates that, under certain conditions, powerful states will cooperate over shared waters.

Dinar (2009: 335–341) provided examples where downstream states have used issue linkage or side payments – a form of compensation offered to upstream states as an incentive to cooperate. Side payments need not be made by states only. In a study of Central Asian waters, Weinthal (2002) showed that transnational actors, such as NGOs and donor agencies, played an important role in providing side payments to the newly independent states in order to encourage their cooperation. These examples show potential strategies that may be used to challenge power asymmetry. The policy literature describes these strategies as developing a 'basket of benefits', which can provide an effective means of enhancing cooperation (WWAP 2003: 319).

Another incentive for cooperation, the spill-over effect between high and low politics, has also been examined. Phillips *et al.* (2006) explored whether cooperation over water issues, or low politics, can foster cooperation in foreign policy and the broader political context, or high politics. Their hypothesis is underpinned by the neo-functionalist IR argument that high politics may be brokered through advancement of low politics cooperation (Mitrany 1975). However, four case studies from the Jordan, Nile, Kagera and Mekong river basins provide inconclusive results on the causal effect between low and high politics. Phillips *et al.* (2006: 186) acknowledged that spill-over is 'a two-way street', where cooperation over water can lead to cooperation on broader issues beyond the river basin and vice versa, but is determined by the basin-specific context.

These studies show that promoting cooperation may be achieved through a variety of means, ranging from economic valuation of water, to institutional design and negotiation strategies. Indeed these aspects are practised, for example, in Southern Africa. As a region with high numbers of shared rivers, the Southern African Development Community (SADC) conducted an economic valuation of water use. It is anticipated that such an assessment would enable a better understanding of optimal allocation and utilization, balancing out sustainability and demands for the resources (SADC 2010). Institutional effectiveness of the RBOs are also sought at the individual basin level within SADC. For example, in the Orange–Senqu River basin, the Water Information System has been set up to enhance data and information sharing. In addition, the RBO has considered the ways in which transboundary environmental assessment could be established with a view to strengthening cooperation (see ORASECOM 2013). However, these interventions do not make power asymmetries between basin states less real or relevant. Decision-making of 'efficient' or 'optimal' water use is still subject to the political will of basin states. Outputs of institutional strengthening in the form of a data portal or a finalized environmental assessment guideline would be the result of a deliberative process. Thus, understanding how cooperation to put in place these measures coexists with conflict is necessary.

Hydropolitics, IR and political science

Hydropolitics has carved out a niche in the examination of issues around water resources management which is distinct from the hydrological and engineering perspectives. This review of studies explaining conflict and cooperation covered a range of theories and hypotheses. In general, studies tend to offer piecemeal insight into why overt conflict does not occur and why non-overt conflict persists. This growing body of literature is not unrelated to the increased interest in sustainable development and environmental security. Sustainable development as a global political agenda has also helped raise the profile of water issues, both in terms of scarcity and watershed degradation (Mirumachi 2015). Research on water and conflict has, in part, been facilitated by the development of environmental security studies that have gained momentum since the end of the Cold War. These studies have examined 'new' threats that the environment poses for the nation-state and those to the environment itself. These new threats necessitated a revision of the role of national security and an examination of the conceptualization of security itself (Ullman 1983; Westing 1986; Mathews 1989; Deudney 1990; Soroos 1994; Renner 1996; Matthew 1997). Within this discussion, natural resource scarcity is often highlighted as a threat to national security. As such, scarcity is considered to be a concern that shapes the debate about the link between lack of water and conflict (e.g. Myers 1989; Homer-Dixon 1991; Gleick 1993; Gleditsch *et al.* 1997). This is not to ignore the fact that disciplines such as geography have made important contributions to the discussion on environmental security (Dalby 1992; Barnett 2001). However, as this review showed, many of the studies on conflict and cooperation over shared waters draw their insight from political science. Consequently, political science, and its subdiscipline IR, has had a strong footprint on hydropolitics.

The analytical lens of Realism, an influential school of thought within IR, is particularly referenced when explaining inter-state conflict and cooperation over shared river basins. Realists argue that cooperation over water is hindered by security concerns, the state preference to reduce interdependence with other basin states, and the tendency to view cooperation as a zero-sum game, or one of relative gains (Dinar and Dinar 2000; Dinar 2009). When cooperation does occur, powerful basin states play a significant role in establishing rules and agreements. In an attempt to develop an analytical framework, Lowi (1993) used realist and neo-realist assumptions to explain when cooperation occurs between upstream and downstream states within a relationship characterized by power asymmetry. Her proposed Hegemonic Theory of Cooperation claims that the downstream state with hegemonic power can bring about cooperation, especially when water is perceived to be vital to its national security. The downstream hegemon uses 'influence, force, "rank", technology, or simply … fear' to secure compliance from the weaker state (ibid.: 169). Her study identified the key analytical components required to provide an understanding of cooperation (or non-cooperation). They are 'resource need/dependence'; 'relative power resources'; 'character of riparian relations or the impact of the larger political conflict'; and 'efforts at conflict resolutions and third party involvement' (ibid.: 11).

Dinar (2009) argued that the IR thinking on institutionalism, rather than Realism, provides a more persuasive explanation of hydropolitics, considering the many historic agreements there have been over water. He claimed that cooperation is the preferred state option, because it is more feasible and effective to set up projects to address issues of water scarcity than engaging in conflict. Factors that explain cooperation, such as the regimes and institutions highlighted above, aim to set out incentives and to achieve compliance. Using the Columbia and Rhine rivers as examples, Dinar (2009) demonstrated how interdependence between basin states enables conditions for both upstream and downstream states to benefit from cooperation. Developing the idea of regimes further, Marty (2001) presented a theoretical model for the assessment of the effective management of international transboundary river basins. Specifically, he focused on the effectiveness of regimes in solving problems that can be analysed by three components. The first component is the nature of the problem. Depending on whether it is a collective problem or an externality problem, the incentives and interests of the states involved will differ (ibid.: 35–38). The second component relates to the strategies and mechanisms, which influence the creation of incentives and the transaction costs for regime formulation and problem-solving (ibid.: 38–45). The third component is identified as a suite of normative criteria for 'good' regimes evaluated on specificity, feasibility, flexibility, organizational features and openness (ibid.: 45–49).

While there is no definitive agreement over the power of explanation between these schools of thought, political scientists Bernauer and Kalbhenn (2010) posited that the major advance in the hydropolitics literature has been the development of quantitative studies that test hypotheses against water wars. They identified two broad approaches of efforts to develop large-*n* studies. The first involves refining models that examine armed conflict and severity of water scarcity; the extent to which the river basin is shared; and the existence of freshwater agreements. Their review found that this approach has enabled more detailed investigation into the causes of conflict and the possible means of resolving disputes, such as those about water quantity and quality (ibid.: 5804–5805). Research by Toset *et al.* (2000), Gleditsch *et al.* (2006) and Hensel *et al.* (2006) demonstrates the development and use of datasets on variables such as the geography of the river and the nature of claims about its use.[3] The second approach assesses the propensity for water wars, through the identification of events relating to water conflict and cooperation (Bernauer and Kalhbenn 2010). This approach is evaluated as instrumental to debunk the water wars myth, providing evidence that militarized conflict between states over water is rare (ibid.: 5805). Most notable has been the development of the Transboundary Freshwater Dispute Database (TFDD), and resultant major publications (e.g. Giordano *et al.* 2002, 2005, 2014; Wolf *et al.* 2003; Yoffe *et al.* 2003).

However, these political science approaches have attracted criticism that they are too state-centric and out of touch with the very real socio-ecological impact at the local scale (Furlong 2006) – as mentioned in Chapter 1. Feminist critiques of this dominant IR approach also draw attention to issues of state-centrism. Detraz (2009: 356–362) argued that the management of the Ganges-Brahmaputra-Meghna River basin is often characterized as one of conflict between basin states, and of environmental security concerns. She criticizes that such analysis is blind to the gendered experience of water resource access and allocation that occurs at the local level. Her study points to an important gap that IR approaches on hydropolitics have not addressed well, noting that 'analysis [ought to] shift away from an almost exclusive focus on the level of the state' (ibid.: 359). Earle and Bazilli (2013) argued that IR is a masculinized discipline and this characteristic (also common to political science) is *not* insignificant. Analysis of hydropolitics drawing on IR is largely dominated by men who influence the research agenda of transboundary water resource management. Consequently, such research does little to change the practice of transboundary water resources management (ibid.). It seems that hydropolitics has been rather insular in developing alternative perspectives, especially when compared with research on sub-national water conflict and cooperation. Politics over water at the sub-national and local levels have been advanced through studies on understanding gendered issues of water access and rights, such as Cleaver and Elson (1995), Zwarteveen (1997), and Crow and Sultana (2002).

Moreover, when closely examining the existing literature employing political science perspectives, it is evident that cooperation is not well defined or explained. Acute conflict over water is relatively easy to identify and define: by observing the existence of armed warfare. Less overt forms of conflict such as diplomatic tension and negotiation disputes rely on analysis that is contextually derived. It is recognized that cooperation cannot be merely analysed by inversing the factors that explain conflict. However, large-*n* quantitative studies, despite their innovation, have yet to yield definitive findings on why and how cooperation develops between basin states, as Bernauer and Kalhbenn (2010) acknowledged. Qualitative studies, on the other hand, do engage with the substance and effectiveness of cooperation, but they have generally failed to employ a means of assessment that is systematic and able to establish causal relationships (ibid.).

To exemplify, particular analytical emphasis defines cooperation by different terms, ranging from an economic perspective (Dombrowsky 2007); from a wider institutional perspective (Marty 2001); or from a list of factors that incorporate the political economy of water, geopolitics, domestic policy-making and international relations (LeMarquand 1977). Other studies provide various terms to emphasize nuances: 'coerced cooperation' (Weinthal 2002); 'tactical functional cooperation' (Sosland 2007); and 'unstable cooperation' (Zawahri 2008). These terms are all useful, in that they communicate that violence does *not* exist and that, instead, various shades of cooperation may reside in transboundary water management. However, it cannot be denied that a robust analytical framework is lacking.

Problems with conceptualizations of conflict and cooperation

This limitation of an abstract understanding of cooperation underpins one of the key arguments I put forward in this book: conflict and cooperation need to be better understood within an original analytical framework that reconceptualizes the relationship between the two. Conflict and cooperation have been characterized as two ends of a polar scale in the existing literature. As critiqued by Zeitoun and Mirumachi (2008: 301–303), the default analytical approach is to use continua that show degrees of conflict and cooperation. The significant development of water event analysis by the aforementioned TFDD is facilitated by an analytical tool: the Water Events Intensity Scale (Yoffe *et al.* 2003). This scale evaluates inter-state occurrences over water within the 15-point range of 'formal declaration of war' to 'voluntary unification into one nation' (ibid.: 1112).[4] Policy documents such as those published by the US Army Corps of Engineers and UNESCO also draw on a continuum of conflict and cooperation, emphasizing how water conflict can be transformed into cooperation. Sadoff and Grey (2005) developed the Cooperation Continuum, in which basin states have been conceptualized to move away from a situation of dispute towards integration by realizing several types of benefits. However, the focus on this linear scale tends to characterize basins in a simplified fashion, in one of two ways: (1) basins in a situation of either conflict *or* cooperation, or (2) basins in a situation that is more conflictual than cooperative/more cooperative than conflictual (Zeitoun and

Mirumachi 2008). For example, Zeitoun and Warner (2006: 443) conceptualized the development of riparian relationships 'as lying somewhere between the extremes of genuine cooperation and cut-throat competition'. Zawahri (2008: 289) noted that studies have focused on identifying and defining conflict and cooperation as opposing concepts, but have resulted in conceptualizations of conflict that is 'simply too wide and general'. She argued that a third category of 'unstable cooperation', which exists between conflict and cooperation, provides more analytical detail. While her focus on a nuanced understanding of conflict is important, the categories, none the less, maintain the linear, scalar conceptualization in which changes may be identified among conflict, unstable cooperation and cooperation.

The problem with this characterization of conflict and cooperation on a linear scale is that it risks analysis becoming apolitical. It strips away the messiness and wickedness of water issues. Only snapshots of transboundary water interactions are picked up. Without understanding the backdrop against which these snapshots occur, there are dangers of developing a superficial if not homogenized view of the causes and consequences. The linear scale with its polarized positioning of conflict and cooperation simplifies complex relations and misrepresents the political context in which riparian states operate (Zeitoun and Mirumachi 2008: 302). Conflict and cooperation on either end of a polar scale presume a linear reality in which states move either towards or away from conflict/cooperation. Aggregating political regimes, hydrology, economic conditions and cultural norms of water into a single, incremental scale is unrealistic. The development of hydraulic infrastructure, the occurrence of floods and the establishment of bilateral/multilateral agreements within a basin do not occur one at a time in a linear fashion which can be scaled as being associated with more cooperative or more conflictual conditions.

Conceptualizing conflict and cooperation on a linear scale is also evident in policy and brings about a tendency to overlook the social and economic capacities required to manage water. The assumptions of water scarcity leading to conflict have shaped policy studies based on water scarcity indices. The Water Stress Index, by Falkenmark et al. (1989), was developed to show that an increasing population could lead to social unrest if the minimum required water supply was not met. But this index gives a false impression that the problem of water supply need only be dealt with to avert conflict. Other water scarcity indices, such as the Water Resources Vulnerability Index (Raskin et al. 1997); the definitions of physical and economic water scarcity (Seckler et al. 1998); and the Social Water Stress Index (Ohlsson 1999), do consider the socio-economic factors of water scarcity, making scarcity less of a technical issue of water supply. However, the Water Resources Vulnerability Index and the physical and economic water scarcity focus on quantitative water availability and do not emphasize the adaptive ability of countries to manage water supply and demand, and to adjust their economies accordingly (Chenoweth 2008: 7).

Analysing transboundary water interaction

So how can we better understand non-linear, complex decision-making over transboundary waters? Suhardiman and Giordano (2012) noted the merit of analysing processes of decision-making that determine hydropolitics and transboundary water governance, so avoiding narrow state-centric analysis. This emphasis on process is beneficial, as it helps embed conflict and cooperation over water in a broader political context. An analytical framework that examines the messy and complex process of conflict and cooperation is needed. Here we turn to studies beyond the political science and IR disciplines for some useful insight.

Polycentric governance, drawn from institutional economics and work on commons by Ostrom (1990, 1999, 2010), attempts to demonstrate the multiple decision-making processes that involve actors beyond the state. This theory has been applied to demonstrate the role and relevance of civil society in transboundary water governance. For example, Cosens and Williams (2012) showed that in order to deal with the complexity and changing nature of problems within the Columbia River shared between the USA and Canada, polycentric governance was a necessary component to the management of resources. They emphasized that local knowledge and capacity is crucial for decision-making in order to make governance effective. In the context of the Santa Cruz aquifer, a transboundary water body between the USA and Mexico, Milman and Scott (2010) found that the reality of institution-building within both countries was polycentric, with multiple agencies involved in the management of the aquifer and environmental affairs at the national and local scales. Importantly, they found that the polycentric nature of domestic decision-making determines the mode and extent of bilateral cooperation. They also pointed out that polycentrism has its pitfalls when issues fall between the jurisdiction and authority of the contributing agencies, thus undermining the capacity of the state to deliberate and negotiate bilateral arrangements. Using a case study of the Syr Darya basin, Wegerich *et al.* (2012) argued that the polycentric nature of decision-making makes international cooperation more suited for the management of tributaries rather than at a basin-wide scale. Their work suggests that multilateral cooperation for an entire basin may only work if agreements are general and broad enough for all parties to agree.

These studies support the relevance of polycentric governance in the empirical context, but theorizing of transboundary water conflict and cooperation is still at the early stages. The study by Myint (2012) attempts to develop a cohesive framework drawing upon polycentric governance. Based on the assumption that problem-solving is not only the purview of state actors, he applies the issues, interests and actor network (IAN) framework. He shows how local communities and NGOs in the Mekong and Rhine river basins have contributed to making rules of water resources management, pointing to the diversity in institutional development. The conceptual approach of polycentric governance provides us with hints on how spatial scales may be incorporated through the way in which actors operate. It is interesting to note that work on polycentric governance has been advanced in the sub-national context (e.g. Huitema *et al.* 2009; Lankford and Hepworth 2010) rather than in the international transboundary water context.[5] This point further adds to the critique that transboundary water politics is theoretically underdeveloped.

One of the early studies attempting to develop an explanatory framework of conflict and cooperation in international transboundary river basins is by LeMarquand (1977). This seminal study focused on three central concepts: 'the hydrologic-economic patterns of incentives and disincentives, foreign policy considerations, and intra-national organizational and regional factors' (LeMarquand 1977: 7). According to this framework, incentives/disincentives of cooperation or conflict are determined by how a state is affected by water use within the basin. This means that water utilization offers different externalities for different states, based on their riparian position. Externalities are mediated through foreign policy that engages with international law and negotiation over the responsibility for, and benefits of, river basin development and management. Basin states may seek to maintain or improve their international image and sovereignty, or to explore options for issue linkage between water and other foreign policy issues. Importantly, LeMarquand (1977: 7–19) pointed out that states take into consideration not only international policy but also the impact and consequences of decision-making at the domestic level. Although LeMarquand's study is criticized as being a 'rather eclectic explanatory framework' (Bernauer 2002: 17), it does provide a useful reminder to examine international and domestic politics, where decisions are made to supply, use, secure and develop shared waters.[6]

Dore *et al.* (2012) provide an inductive framework that explores the decision-making process and context, drawing upon empirical insight from the Mekong River basin. This 'heuristic framework' examines how multiple stakeholder decision-making may be expressed as comprising drivers and arenas (ibid.: 23). Drivers are made up of the interests of the various actors involved and the discourses they use to frame issues and establish institutions. These drivers play out in arenas, or decision-making processes, that are fundamentally political and mediated through the actors exercising power. A useful point to take away from this framework is how the interplay between actors, and the tools employed to influence decision-making, do not occur within a vacuum but rather in a physical and socio-economic context, specific to the river basin concerned; in other words, decision outcomes and their impact as being contingent to the characteristics of the decision-making arena. This framework directs examination of the way in which transboundary water governance is shaped, beyond specific drivers of conflict and cooperation. A weakness of this framework is that the ontological assumptions about actor interests and power are not explicit, and the facilitating methodology is absent from the discussion. None the less, this study is another useful example that emphasizes the process involved in transboundary water governance, demonstrating how the literature on hydropolitics has advanced beyond a narrow disciplinary focus based on environmental security.

While none of the analytical frameworks highlighted above is perfect, they do identify some important aspects about the makings of hydropolitics, and the conflict and cooperation associated with it. If we are to examine the political process in which conflict and cooperation occurs, rather than cooperative events or conflictive outcomes, spatial scale must be an important consideration. Studies on transboundary water governance indicate that state and non-state actors play a role in decision-making, at least theoretically. The implication is that the actors involved in decision-making cannot be homogenized, and that more attention needs to be paid to the power relations of these actors.

In addition, these actors are dispersed across various spatial scales, or as Mollinga (2008a, 2008b) describes them, as 'domains' of politics. Global politics, inter-state hydropolitics, politics of water policy and everyday politics are the domains in which actors engage in decision-making about water. Decisions in one domain inevitably have implications for another, because of the way policy and contestations over water occur. This does not necessarily happen top-down (i.e. global to local domain), but sometimes in unexpected ways (ibid.). Consequently, it becomes necessary to understand how decision-making at one spatial scale is influenced by other scales.

I argue that transboundary water interaction is central to the analysis of transboundary water politics, underpinned by a culmination of decisions at different spatial scales. Interaction among stakeholders that sustains management and governance of the shared rivers, rather than individual conflictual or cooperative events, should be the primary focus of analysis (Mirumachi 2007a; Zeitoun and Mirumachi 2008). By understanding how interaction changes, a wider range of drivers of conflict and cooperation may be identified. Water resources use is embedded in the socio-economic conditions of a basin, and modifying water-use patterns is not simple. The management and governance of shared basins need to contend with factors outside the 'water box', such as those relating to land use, food security and energy demands. In other words, the explanatory factors of transboundary water interaction changes are not necessarily found in one spatial scale. This means that it is necessary to recognize that decision-making outside of this water box is fundamental to the outcome of river basin development (WWAP 2009). Conflict and cooperation over shared waters are thus also subject to broader political contexts and the political economy of water.

Conflict, cooperation and power

A focus on interaction influenced by scalar linkages means, analytically, distancing oneself from normative assumptions of conflict and cooperation. In other words, whether the process of river basin planning is beneficial or controversial depends upon who is involved in the interaction, and the global, domestic, local politics that shape this inter-state interaction. There is recognition in scholarly literature that cooperation and conflict are multi-faceted. For example, Robert Axelrod, the leading thinker on game theory, highlighted the fact that cooperation is assessed subjectively; for some it may be good, for others bad:

> Cooperation need not be considered desirable from the point of view of the rest of the world. There are times when one wants to retard, rather than foster, cooperation between players. Collusive business practices are good for the businesses involved but not so good for the rest of society. In fact, most forms of corruption are welcome instances of cooperation for the participants but are unwelcome to everyone else.
>
> (Axelrod 1984: 18)[7]

Frey (1993), who called for better theorizing within hydropolitics, noted the nuanced nature of cooperation. Moreover, he pointed out that conflict should not be assumed to be *bad*:

> Both conflict and cooperation refer to interaction among actors, individual or group, that is, to systems of interactor influence or power. Conflict is by no means always or inherently negative, either for the individual actor or the social system.... Nor is cooperation always positive, as, for example, in situations of collusion.
>
> (Frey 1993: 57)[8]

It has also been argued that tension can clarify the diversity of contentious issues and bring about change. When considering the way in which policy frameworks to manage water and land could be strengthened, Roberts and Finnegan (2013: 5) asserted: 'Conflict in itself is not negative. It is an inevitable part of life and can function as a motor for change and development in society if handled constructively.' These functions of conflict may be seen in multi-stakeholder platforms. These deliberative tools offer a discursive space where multiple views and values invested in water resource management can be 'fought' out by different parties. At the same time, these political arenas can offer opportunities for the resolution of issues that may not have been possible elsewhere (Warner 2005).

However, as discussed earlier, the tendency of hydropolitics literature is to conceptualize conflict and cooperation on a linear scale. Such analysis judges conflict as the undesirable outcome of interaction between riparian states and cooperation as the desirable outcome (Zeitoun and Mirumachi 2008). These notions of conflict and cooperation are problematic because there are cases where cooperation may not necessarily solve problems within the basin. Normative assumptions can lead to a situation where:

the mere existence of cooperative arrangements is celebrated as a sign of progress, with little or no interrogation of their influence and impact in addressing the core problems presented by trans-boundary waters.

(Kistin and Phillips 2007: 2)

Cooperation does not necessarily lead to equitable water allocation outcomes, meaning that 'not all cooperation is pretty' (Zeitoun and Mirumachi 2008: 305–306). Selby's (2003) critical analysis shows that Israel takes the approach to 'dressing up domination as "cooperation"' over issues with Palestine of water rights and allocation. Consequently there has been little change to water allocation in this region. 'Cooperation' in such guise may suit the elite decision-makers in Israel and Palestine, but not others – to draw upon Axelrod's observation. While there may be a joint organization, the Israeli–Palestinian Joint Water Committee, water allocation remains highly skewed, prompting questions about the extent to which cooperation in the form of bilateral institutions can challenge the status quo (Zeitoun and Mirumachi 2008).

Here, an explanatory analysis of power asymmetries between basin states is useful. Power asymmetry not only explains how conflict about water allocation and exploitation of resources occurs, but also how consent in the form of agreements and institutions may be established. In particular, the FHH explained earlier is informative, because it suggests different dimensions of power that may be used to secure water resources (Zeitoun and Warner 2006). Rather than using brute power, or the power from superior military and economic strength, it may be that bargaining or ideational power is effective in bringing about outcomes that are not contested, under the guise of cooperation. Bargaining power may be utilized in negotiations to offer incentives and trade-offs that lead to some form of consensus between basin states. Ideational power is about 'power-over-ideas' (ibid.: 443). This dimension of power is considered to be the most effective in shaping water allocation outcome, because it influences how actors believe and perceive issues (Zeitoun and Allan 2008). Once mindsets are conditioned to view a particular river basin management as rational or fair, then overt and coercive measures for compliance are no longer necessary. Thus, a situation arises where inequitable or unsustainable water allocation arrangements are not disputed, and instead seem to represent cooperative riparian relationships (Zeitoun *et al.* 2011).

Employing the lens of power can provide a critical analysis of what Dinar (2009) called 'benevolent hegemons'. He claimed that downstream states, such as India and Egypt, act as benevolent hegemons to their respective upstream basin states, offering economic incentives designed to realize cooperation (ibid.: 341–352). For example, when negotiating the *1959 Agreement for the Full Utilization of the Nile Waters* with Sudan, Egypt made certain concessions on water allocation and compensation arrangements. It is argued that 'India has taken it upon itself to provide all the financing for the project' in Bhutan to realize bilateral hydropower trade (ibid.: 350). The opportunity presented by Indian financing is described as progressing development initiatives for the weaker upstream state, beyond mere financial return from hydropower sale. However, by examining in detail how other forms of power are used, and in what kind of wider political context, a different evaluation may be given to the actions of Egypt and India. Cascão (2009a) provides a critical analysis of how the *1959 Agreement* only served to further Egypt's claim to a historic right to water allocation when Sudan (and other countries) had very few means to challenge it. The consequence of this being asymmetrical control of water resources was further entrenched. For the India–Bhutan case study, the nuance may not be so much that India benevolently took on costs for hydropower development. Rather, India has used its power, both economic and ideational, to convince and comply with neighbouring states in bilateral projects – as will be shown in Chapter 4 between India and Nepal. In other words, understanding the quality of consent provided by the weaker upstream states is pertinent for a critical analysis. This is especially important when these states do not see the hegemon as being particularly benevolent and have no choice but to accept what is being offered. Overlooking the subjective nature of conflict and cooperation misses the point on the political nature of transboundary water interaction.

Coexisting conflict and cooperation

The linear conceptualization of conflict and cooperation over international transboundary river basins falls into the trap of binarism. The analysis of water resources governance often uses binaries such as 'nature–society, global–local, modern–traditional, and state–village' (Mollinga 2010: 423). I argue that conflict–cooperation is another of these binaries that has coloured the way in which analysts understand international transboundary waters. These binaries are analytically problematic because they fundamentally ignore the complex reality of water resource management and present instead a reductionist version as reality (Mollinga 2010). This prevents analysts from understanding the multiple faces of both conflict and cooperation. Thus, it is imperative to examine conflict and cooperation together if we are to make sense of transboundary water interaction about allocation, river basin planning and watershed management. The politics of international transboundary river basins is about both the conflict and cooperation of actors, and the dynamic process in which they engage. Studies that pose research questions on *either* conflict *or* cooperation have epistemological problems to explain in full transboundary water politics. This binarism needs to be treated with caution, not just in the analysis of transboundary water politics but also water resources management issues at other spatial scales.

Some studies identify how conflict and cooperation occur at the same time. For example, Daoudy's work (2005, 2009) based on negotiation theory examines both conflict and cooperation during negotiations about water resources allocation. She posited that negotiation processes include both distributive approaches – with gains for only one actor – and integrative approaches – with the potential for mutual gain. Her empirical analysis on relationships between Syria, Turkey and Iraq found that they were characterized by 'peaks of conflict *and* periods of mutual cooperation' (Daoudy 2009: 364; emphasis in original). However, I argue that this analysis is more useful for demonstrating the interchangeable use of distributive (conflictual) strategies or integrative (cooperative) strategies employed by actors during the negotiation process, rather than rejecting binaries altogether. In the Incomati River basin, van der Zaag and Carmo Vaz (2003) analysed that there has been both conflict and cooperation over the issue of abstraction between Mozambique and South Africa, resulting from a mix of a shared cultural heritage, institutional development and engineering options to increase water allocation. Some cursory recognition is provided by Zawahri and Gerlak (2009: 218) that both conflict and cooperation need to be part of the research design when studying international transboundary waters. However, there is little indication about how to do this, and overall, the scholarship has yet to develop an alternative approach that does not resort to simple binaries.

Conclusion

The inadequacy of the current literature presents us with the challenge to explore transboundary water interaction that is characterized by both conflict *and* cooperation. Examining conflict and cooperation separately is simply not sufficient and even risks a detrimental policy outcome. Policy discussion has tended to focus only on cooperation without addressing conflict (Zeitoun and Mirumachi 2008). For example, Green Cross International (2000) categorized cooperation over international transboundary rivers as representations of allocation, salvation and opportunity. In this way, positive normative values are attributed to cooperation. This approach over-simplifies the problem of water resource management, especially in asymmetrical power contexts, and ignores the political sensitivity of water allocation and use (Zeitoun and Mirumachi 2008).[9] Moreover, if policy ignores or undermines the political context in which international transboundary interaction occurs, it is prone to the perpetuation of sub-optimal, inequitable and unsustainable water use and allocation (ibid.: 306–307).

A review of the existing literature reveals different ontological emphases and epistemological accounts of the way in which conflict and cooperation over shared waters are understood. The various case studies, models and frameworks examined in this chapter point to the fact that political analysis cannot be ignored. However, despite concerns about the so-called global water crisis, it seems that scholarly work is not well equipped with an analytical framework to understand conflict and cooperation among basin states without falling into the trap of binarism. In the next chapter, I will develop an original conceptual framework to *simultaneously* understand riparian interactions characterized by coexisting conflict and cooperation intensities.

Notes

1 Other factors pointed out by Schmeier (2013) include how RBOs collect and manage information and data, monitor environmental conditions within the basin, and resolve disputes between basin states.
2 See studies on negotiation theory by Lax and Sebenius (1986) and Sebenius (1992) that explain the notion of 'zone of possible agreement'.
3 See Shared River Basin Database and other related datasets available through the Peace Research Institute Oslo (PRIO) (www.prio.no/Projects/Project/?x=724), and the ICOW River Claims Dataset from the Issue Correlates of War (ICOW) project (www.paulhensel.org/icow.html).
4 In a later publication by De Stefano *et al.* (2010) which updates findings of conflict and cooperation using the TFDD, the notion of coexisting conflict and cooperation is accepted after Zeitoun and Mirumachi (2008), and results show how basins are in flux in their patterns of conflict and cooperation.
5 See also Schnurr (2008) and Pahl-Wostl (2009) as examples of emerging work on the global governance of water resources from a polycentric perspective.
6 Refer to studies by Allouche (2005) and Sebastian (2008) for the application of LeMarquand's framework for river basin management with regard to the Indus, Jordan, Orange–Senqu and Okavango river basins, and the Aral Sea region.
7 See also Axelrod and Keohane (1985: 226).
8 Frey (1993) none the less conceptualized conflict and cooperation as opposites. In his view, cooperation occurs when common goals are achieved and conflict occurs when goals are disputed (Frey 1993: 57).
9 See also UNDP (2006: 228) for another example, which focuses solely on achieving cooperation.

3 The Transboundary Waters Interaction NexuS (TWINS) framework to understand coexisting conflict and cooperation

Introduction

The review of the literature on conflict and cooperation over international transboundary waters reveals that a study of just conflict *or* cooperation offers only a partial picture of riparian relationships. This serious limitation provides a basis for the research of this study. It is argued that the focus of analysis should be on riparian interaction as a process that involves a range of actors, rather than on outcomes of conflict *or* cooperation. The study posits that conflict and cooperation coexist. Transboundary water interactions inherently involve both conflict and cooperation, and a study of this process must necessarily examine the complex politics in which shared water resources are controlled, negotiated and governed. A comprehensive conceptual framework, which analyses conflict and cooperation simultaneously, is imperative for this purpose. This chapter sets out to develop a conceptual framework of transboundary water interaction informed by interdisciplinary insight into conflict and cooperation.

The Transboundary Waters Interaction NexuS (TWINS) is presented as a robust framework with which to examine coexisting conflict and cooperation through the evolution of relationships between basin states. A key analytical tool within this approach is a matrix containing different combinations of conflict and cooperation intensity. This matrix helps illustrate the degree of change to coexisting conflict and cooperation. Power analysis offers useful explanations for such changes observed through the use of the matrix. The following sections detail the TWINS framework and its advantages; the matrix of coexisting conflict and cooperation; and key concepts that inform analysis within the framework.

The TWINS framework for examining coexisting conflict and cooperation

Various theories that deal with environmental issues, such as conflict resolution, negotiation and political psychology, refer implicitly to conflict and cooperation as being interrelated (Zeitoun and Mirumachi 2008: 299). Political science and IR, disciplines from which many of the studies reviewed in the previous chapter originate, also recognize the nuanced nature of conflict and cooperation among states. For example, McMillan (1997: 40) noted that 'States may engage in conflict and cooperation at the same time, and interdependence may be related to both outcomes'. And Vasquez (1995: 138) usefully pointed out that '[c]onflict can occur within the context of a cooperative relationship, and cooperation can occur within the context of conflict'. IR scholars Axelrod and Keohane (1985: 226) indicated that 'cooperation can *only* take place in situations that contain a mixture of conflicting and complementary interests' (emphasis added). This is a highly significant observation. However, 'mixtures' need to be understood via an alternative approach to the linear scale, as argued in Chapter 2, but an explicit framework to simultaneously examine conflict and cooperation is lacking within existing hydropolitical studies.

In order to overcome this limitation, this study uses an interdisciplinary approach. Here I draw ideas from Craig (1993), whose sociological work on cooperation informed that normative assumptions about 'bad' conflict and 'good' cooperation may not be helpful:

> Conflict is a concept that is independent of co-operation; not always opposite to it. In certain circumstances, conflict may be an integral part of inducing and sustaining co-operative behaviour, and the two may coexist in various social settings.
>
> (Craig 1993: 15)

Craig (1993) devised a simple but very effective matrix that provides understanding of how different intensities of conflict and cooperation can coexist (see Figure 3.1). His two-by-two matrix of high and low conflict and cooperation is a powerful tool to explain the differing nature of relationships. Specifically, where there is low conflict and low cooperation, there is little interaction among social actors. Where high conflict and low cooperation coexist, the nature of the relationship is unstable. On the other hand, where low conflict and high cooperation coexist, the relationship is stable. High conflict and high cooperation give rise to a situation where '[t]here is strong commitment to achieve a goal by participants, but there may be equally strong disagreement over the precise definition of that goal and particularly over the means of achieving it' (ibid.: 16).

The recognition that conflict and cooperation coexist is intuitive, and perhaps best characterized by interpersonal relationships such as marriage or any family relationship where consensus, compromise and tension exist side by side.[1] The matrix of conflict and cooperation is useful, as it sets us on a course to analyse different combinations and their implications as to the nature of relationships (for example, those with little interaction; those which are stable; and those which are unstable). Craig's (1993) persuasive ideas on the coexistence of conflict and cooperation can be scaled up and applied

		Cooperation	
		Low	High
Conflict	Low	Little interaction	Stable and comfortable
	High	Unstable relations	Unstable, intense, sometimes creative

Figure 3.1 Coexisting conflict and cooperation (Source: adapted from Craig (1993: 16). Permission granted by Black Rose Books, Montreal, www.blackrosebooks.com).

beyond interpersonal relationships to formal transboundary water relationships, negotiated by representatives of basin states.

This study presents the Transboundary Waters Interaction NexuS (TWINS) framework in which a matrix of coexisting conflict and cooperation is a key component. The original two-by-two matrix by Craig is developed further with multiple levels of intensity to highlight nuanced transboundary water interactions (see Figure 3.2). With the matrix representing a trajectory of changing coexisting conflict and cooperation, it is possible to use it to examine the factors contributing to the nature of international transboundary water interactions. The TWINS framework is a way of examining the development of basin relationships, using a matrix of conflict and cooperation intensity, modified for the hydropolitics context.[2] The framework assumes that transboundary water interactions are not static but rather in constant flux, influenced by, and influencing, the broader political context in which they occur.

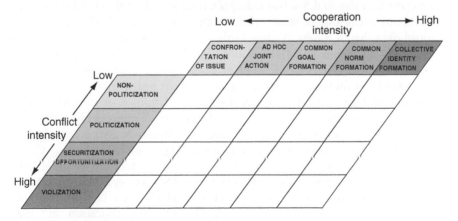

Figure 3.2 The TWINS matrix.

Using the TWINS matrix as an analytical tool, this study is informed by constructivist assumptions about social order and agency. While existing studies tended to view conflict and cooperation over international transboundary waters from a realist or neoliberal institutionalist perspective (e.g. Lowi 1993; see also Chapter 2), this study takes a different tack, finding the constructivist perspective more useful. Different actors engaged in water resources management establish their identities through engagement with other actors. Moreover, their identities cannot be assumed to be a given and may change (Wendt 1999). To exemplify: the ministry of water affairs of a basin state may establish their identity as a regulating agency through confrontation over water scarcity; an irrigation board made up of local farmers may establish their identity as a water management institution in its interaction with other local water users. Order within a basin, exemplified in RBOs or through the practice of international water law, is based on socially constructed meanings of governance and sustainability that are deliberated and contested by actors involved.

An analysis of coexisting conflict and cooperation unveils the complex, intersubjective political process of riparian interaction. This approach helps bring nuance into understanding transboundary water interaction, rectifying the over-simplistic characterization of basins as being either in conflict or subject to cooperation. In addition, this approach forces the analyst to closely examine the changing nature of the degree of coexisting conflict and cooperation, and the reason for this. The relationship between agencies, and the structures, become focal points of analysis in this constructivist approach (Klotz and Lynch 2007).

Examining conflict and cooperation simultaneously, through a matrix, has proven to be useful for a number of studies on transboundary water politics. For example, Sojamo (2008) showed how relationships between Uzbekistan and its neighbouring states – Kyrgyzstan, Kazakhstan, Tajikistan and Turkmenistan – showed different, and changing, levels of coexisting conflict and cooperation in the Aral Sea basin. Importantly, she identified the fluid nature of bilateral relationships which reflects the struggle for hydro-hegemonic control over shared waters. Adding further detailed insight on power relations is the study by Menga (2014), which uses the TWINS matrix to analyse how Uzbekistan's hydro-hegemonic control over water resources is increasingly challenged by Tajikistan and Kyrgyzstan in recent years. Increase in conflict intensities in Uzbek–Tajik and Uzbek–Kyrgz relations is attributed to tensions over the Rogun and Kambara dam development respectively, which are promoted in a variety of ways by the non-hydro-hegemons, including rallying for international support and mobilizing financial resources. Conti (2014) plugged an important gap in the hydropolitics literature by investigating transboundary aquifers, drawing on the TWINS concept. Her study helps to provide an overview of the level of reliance which different states place on aquifers, and the different issues associated with their allocation and use. Using the TWINS matrix, Warner and van Buuren (2009) examined how not only state actors, but also non-state actors, can be key agents of complex decision-making with regard to water. They found that the Scheldt River shared among the Netherlands, France and Belgium brought together not only Dutch and Flemish state actors but also business and NGO actors. Their findings demonstrate how deliberative processes do not eradicate conflict, but rather present opportunities for conflictual strategies:

> Especially when actors hesitated to participate in a collaborative process, the opportunity to use conflictual strategies alongside talk and deliberation motivated them to participate in collaboration. They then could put pressure on the process when they felt threatened. The knowledge that there is always the possibility of bypassing the collaborative process or even exiting it via a political lobby or a lawsuit convinced actors to participate in a cooperative process because they knew their fallback option.
>
> (Warner and van Buuren 2009: 436)

Their study explains the evolution of decision-making over shared waters – not only through bilateral international diplomacy but also within the domestic politics of the basin states.

Allan, in Allan and Mirumachi (2010), developed another dimension to the TWINS matrix by adding a third axis to measure the level of resource use. This version of the TWINS matrix shows how levels of water resource exploitation and conservation are fundamentally determined by solutions outside the watershed, such as industrialization and the nation-state's diversification of economic activity. This interpretation of the TWINS matrix reminds us that it is not the characteristics of the watershed that we should be preoccupied with but rather the nature of the 'problemshed' (Allan 2001). His interpretation is useful in that it highlights the fact that coexisting conflict and cooperation is subject to broader political and economic structures.[3]

Moreover, the TWINS approach has proven useful with regard to other transboundary environmental issues. The study by Martin et al. (2011) used a modified version of the TWINS matrix to show the evolution of management of a transboundary natural park between Rwanda and the Congo. Their study confirmed that the traditional approach of examining international relations over transboundary environmental problems as either an issue of conflict or cooperation was not sufficient. They pointed out that issues of low politics, in their case the environmental management of the natural park, were enmeshed in high politics, such as regional security. The empirical analysis of this current book, on transboundary water issues in the Orange–Senqu, Ganges and Mekong basins, shows the ways in which low and high politics mix. But as will become clear in the following chapters, I further analyse this relation between low and high politics, and argue that low politics of the environment can also become high politics. The TWINS matrix with tailored conflict and cooperation intensity scales used by Martin et al. (2011) underlines the importance of understanding the broader context in which transboundary environmental issues occur, in order to provide explanations of increased/decreased instances of conflict and cooperation intensity.

Actors and speech acts

Having established the analytical approach and tool to examining both conflict and cooperation simultaneously, I now turn to the identification of key concepts that will situate the analysis of transboundary water interactions. The first concept is the hydrocracy. The hydrocracy, or the hydraulic bureaucracy, comprises bureaucratic agencies, such as ministries and departments responsible for agriculture, irrigation, water resources and energy (Wester 2008). By using the concept of the hydrocracy, the TWINS framework focuses on elite decision-makers who are influential and responsible for actions regarding water allocation and utilization, and transboundary river basin management more generally. The hydrocracy accumulates vast amounts of knowledge and information through their use of technical expertise about potential river development projects that informs the state agenda. In addition, this group of actors is often supported by politicians seeking to influence (and gain) from these projects (Molle *et al.* 2009).

The focus on these state elites who frame water issues is an explicit attempt to uncover the rationale and justifications of engineering the river basin. In this way, the study unpacks the 'state', and avoids homogenizing it in a way that has been critiqued in studies such as Furlong (2006) and Sneddon and Fox (2006). It goes without saying that the framing of water issues is not always exclusively done by elite decision-makers. Non-state actors such as businesses with a stake in water resources development, NGOs that pressure governments for better practice, international funding organizations like the UN and donor agencies and local communities do play a part. However, the make-up of these elite decision-makers enables them to gain access to a number of fora to deliberate and negotiate transboundary water allocation and use at the international level (e.g. through representation in RBOs) and at the national level (e.g. through policy formulation of national water management strategies). How these elite decision-makers enforce decisions, buffer criticism or rally support for river basin development projects reflects the political economy of water allocation and use. Specifying agency in this way provides further insight into the mechanics of state decision-making that can be opaque.

The second concept used in the TWINS framework is speech acts of these state elites. Speech acts are verbal acts that create social facts, establish relations between actors and construct the rules of relationships (Austin 1962; Searle 1969). An example is the oft-cited explanation given by Austin (1962: 5) that a marriage vow is a verbal speech act that establishes the rules of marriage between two individuals. In this way, speech acts represent a constructivist view of how relationships are established. In addition, speech acts may be nonverbal acts that also establish or change the rules between actors (Frederking 2003: 367). For example, deploying a missile is a nonverbal speech act designed to achieve compliance, or to increase political influence over the recipient of the speech act (ibid.: 367; Duffy and Frederking 2009: 328).[4] It is proposed that it is through speech acts that these powerful actors socially construct structures through which shared waters are managed and governed. At the same time, through this process of socialization, the identities and interests of the actors themselves change and develop. In other words, socialization is about both social order and the actors changing through discursive means (Wendt 1999; Klotz and Lynch 2007). Therefore, speech acts are useful indicators of how such change occurs.

Speech acts have the power to assert, direct or commit (Duffy and Frederking 2009: 328). Public statements and declarations can be assertive speech acts that signal reciprocal understanding of an issue between two states. By imposing an economic sanction or effecting military intimidation, states make directive speech acts in order to secure the compliant behaviour of their recipient. International agreements are commissive speech acts, binding the actions of a state to the content of the agreement (ibid.: 328). In the context of international transboundary water resources management, an assertive speech act may be a joint declaration on water quality improvement. A directive speech act may reflect the action of the hydrocracy of an upstream state, which secures its water resources by closing dam gates or unilaterally constructing hydraulic infrastructure, without the consent of other states downstream. Signing a bilateral treaty on water sharing would be an example of a commissive speech act.

Speech acts are used as markers of coexisting conflict and cooperation, and are plotted in the TWINS matrix. By way of explanation, if the directive speech act of unilaterally constructing hydraulic infrastructure is to the detriment of downstream states, conflict intensity would increase, and could prompt changes in cooperation intensity. Control of the river may be deemed necessary by the elite decision-makers of the upstream state, and the construction as a means of ensuring human safety and economic activity. In contrast, a commissive speech act such as signing a water treaty may increase cooperation intensity in inter-state relations.

Commissive speech acts like this become vehicles through which norms and ideas are conveyed, and the basis of cooperative interaction between basin states. In other words, the cooperation intensity employed in the TWINS matrix reflects the process of socialization on issues of water resources management and governance. From a constructivist perspective, the diffusion and adoption of norms and ideas are important to this process (Checkel 1999). For the purpose of this study, norms are defined as 'shared expectations about appropriate behavior held by a community of actors' (Finnemore 1996. 22). In the context of environmental management, norms that are considered 'good' or 'useful' can instigate cooperation when they are collectively understood and adopted in policies (O'Neil et al. 2004: 160). Norms can also be supported by ideas that inform state interests and policy (Ringius 2001). These ideas are defined as public ideas: 'widely accepted ideas about the nature of a societal problem and about the best way to solve it' (ibid.: 1).

Speech acts are observed through a collection of documents; for example: public declarations; minutes; media reports; newspaper articles; legal documents such as treaties and agreements; biographies and memoirs of decision-makers; brochures and newsletters; and policy briefs. Water-specific databases, such as the International Water Events Database of TFDD, and the Water Conflict Chronology of the Pacific Institute, are useful starting points to identify speech acts.[5] However, the additional use of documents, such as minutes and a range of grey literature, is needed to qualify levels of coexisting conflict and cooperation intensity. Document analysis may be combined with data from interviews with members of the hydrocracy and politicians, as well as non-state actors who advise and lobby decision-makers. This process complements the TWINS analysis by triangulating specifics of the speech acts and uncovering further data unavailable in published documents.

In summary, identifying and analysing speech acts of elite decision-makers represent a way of tracing the changes in conflict and cooperation intensity on the TWINS matrix. Different types of speech acts are understood to construct transboundary water interactions. The effect of speech acts can vary: some may increase conflict intensity and decrease cooperation intensity; and others may increase cooperation intensity whilst maintaining conflict intensity levels. In the following two sections, the scales of conflict and cooperation intensity, employed in the TWINS matrix, are explained in more detail.[6]

Conflict intensity

The TWINS matrix uses a scale developed by Warner (2004a, 2004b), specifically devised for transboundary waters, that draws upon the ideas of securitization. There are four levels of conflict intensity within the scale: non-politicized, politicized, securitized/opportunitized and violized (ibid.). This scale reflects the process which fosters the increasing prioritization of issues on political agendas, to the extent that acute militarized action is taken to satisfy them. Securitization theory (Buzan *et al.* 1998) explains the ways in which issues can become important such that extraordinary practices of decision-making are justified.

Issues that do not concern the state, or issues that are not in the public domain, are non-politicized issues. Once an issue enters the political agenda, it is politicized. The issue then becomes 'part of public policy, requiring government decision and resource allocation' (Buzan *et al.* 1998: 23). The implementation of the European Union Water Framework Directive is an example of a politicized issue. The implementation of national water policy in accordance with the directive requires parliamentary decisions, scientific and legal expertise, and other resources to meet water quantity and water quality obligations.

When 'the issue is presented as an existential threat, requiring emergency measures and justifying actions outside the normal bounds of political procedure', it is securitized (Buzan *et al.* 1998: 23–24). While there are different approaches within securitization theory, the Copenhagen school of thought is referenced in the TWINS framework in order to illustrate the social construction of threats and crisis. Threats do not have to be real or already existing to make a securitizing move, but may be discursively framed for the purposes of 'special' politics. As Wæver (1995: 57) claimed, 'The utterance [of security] itself is the act'. In order to make extraordinary practices plausible, there must be a sense of urgency that 'normal politics' cannot deal with sufficiently (Laustsen and Wæver 2000). Thus, threat and urgency are important aspects of securitization. Speech acts have the power to threaten and construct urgency.

Securitizing speech acts '[take] politics beyond the established rules of the game and frames the issue either as a special kind of politics or as above politics' (Buzan *et al.* 1998: 23). An example of a securitizing speech act regarding shared waters is found in Allouche (2005:16). In responding to a water transfer project proposed by Lebanese decision-makers in the Hatzbani River of the Jordan River basin, the Israeli Minister for Infrastructure declared the following: 'Israel cannot let pass [this decision about water transfer] without reaction. For Israel, water is a matter of to be or not to be, to live or to die' (Shuman 2001: n.p.). This claim demonstrates how the planned water diversion by the Lebanon became an existential threat for Israel such that actions to ensure its survival were necessary and urgent. Warner (2004b: 9, citing Warner 2004a) contended that issues may be opportunitized when they 'offer such a chance to improve a situation that it justifies actions outside the normal bounds of political procedure'. As this would require extraordinary measures, Warner equates the conflict intensity with securitization. Opportunitization, as understood by the TWINS framework, relates to the removal of public discussion by recognizing an act of emergency. It should be pointed out that whether an act of emergency leads to an improvement in the situation is subjective. In other words, a securitizing/opportunitizing speech act may improve the situation from the perspective of the actor declaring emergency measures, but not necessarily from the perspective of the recipient of the speech act. A stark example of securitization/opportunitization involving an Indian speech act, which led to the unilateral construction of hydraulic infrastructure during the early 1990s on the Mahakali River, is analysed in Chapter 4.

At the most extreme, interaction goes beyond securitization to being violized (Warner 2004a, 2004b). Drawing on arguments by Neumann (1998), the conflict intensity scale includes instances where violent action is seen as the necessary response (Warner 2004a, 2004b). As Wolf (1998), Homer-Dixon (1999) and Yoffe *et al.* (2003) have shown, violized interaction over water at the international level has been non-existent.

Cooperation intensity

Compared to conflict intensity, cooperation intensity is not well conceptualized in the context of international transboundary waters. The literature on international transboundary waters offers a range of cooperation factors but no holistic analytical framework, as mentioned in the previous chapter. Importantly, Gerlak and Grant (2009: 138) found that:

> no one theoretical [IR] perspective (neorealism, neoliberal institutionalism, or neofunctionalism) fully explains the creation of cooperative institutional arrangements in international river basins, though all make important contributions.

This points to some of the weaknesses within IR with regard to the definition and explanation of cooperation more generally. Game theory has made some headway, best exemplified through the seminar work of Axelrod (1984). His work showed that reciprocity is vital in situations characterized by the Prisoner's Dilemma where actors may mutually gain, mutually lose or unilaterally gain from individual, uncoordinated action. However, game theory assumes there to be a choice between cooperation *or* conflict. Analysis based on this theory risks failing to grasp the *simultaneous* nature of conflict and cooperation in relationships that Craig described (1993). Some scholars have attempted to provide more analytical rigour on cooperation. For example, Zawahri (2008), who rightly argued that existing literature on international transboundary waters could be refined, offered a detailed list of cooperation and conflict definitions based on the nature of problems being negotiated between states. In addition, studies by Marty (2001), Kistin (2010) and Schmeier (2013) focus on institutions of water governance as a proxy for cooperation. However, the literature is scant on a definitive scale of cooperation intensity to examine the politics of transboundary water resources management.

This study argues that cooperation is a reflexive process, in which norms and ideas about water resources management and governance are shared by riparian states. A constructivist understanding assumes that states can develop collective identities that bring about collective action. If a certain norm, such as one of collective security, is internalized, it may lead to the formation of collective identity between states and result in collective action (Wendt 1999: 302–307). This collective identity also results in international interests becoming part of national interests (ibid.: 304–305). Hence, the process of cooperation has a similar effect to that of securitization, in that it creates specific social realities. Warner, in Mirumachi and Warner (2008), argued that cooperative speech acts, such as those that develop partnerships, have the effect of coordinating interests and sharing common goals.

Following this, the cooperation intensity of the TWINS matrix is based on the extent to which there is intention for collective action and common goals, operationalized through the identification of norms and ideas. Frey

(1993: 57) understood cooperation as the 'coordination of behavior among actors to realize at least some common goals'. Common goals are required for cooperation, and coordination of behaviour refers to joint action. In addition, Frey argued that actors strive for a common goal because of the *'felt* agreement or compatibility of goals' (ibid.: 57; emphasis added). This implies the intention of collective action. This aspect of intent is important, and we can draw on some useful insights from Tuomela (2000). Coming from a philosophical perspective, he defined collective action as not just about joint action but also about the mutual belief in collective reasons based on common goals and practices. His argument is that collective action brings about dependence, which is 'collectively intentionally (hence voluntarily)' formed (ibid.: 73).[7]

The intent to engage in collective action, and the acknowledgement of common goals, is underpinned by water management norms and public ideas. For example, norms established through international water law, such as the obligation to cause no harm to other states through one's own water use, can inform the practice of water resources management. As the case studies to be viewed later demonstrate, harnessing rivers for economic development is an example of a public idea which was advocated in the Orange–Senqu, Mekong and Ganges River basin states. This public idea provided a way to argue the need to prevent flooding and water scarcity, often by damming the river. In this way, these norms and ideas constitute the tools that shape the actions of states.

Based on these elements of intention for collective action, common goals, norms and public ideas, five indicators of cooperation intensity are used in the TWINS matrix. They are: confrontation of issue; ad hoc joint action; common goal formation; common norm formation; and collective identity formation. As mentioned earlier, it is difficult to pin down the meaning of 'cooperation'. For this reason the cooperation intensity scale is presented as one interpretation of tracing the social construction of relationships, rather than a definitive scale.

In bilateral riparian relationships, the issue of transboundary water resources management may be of concern to both states. Action taken in domestic politics demonstrates acknowledgement of the issue as a problem. The first level of cooperation intensity is such confrontation of the issue. In these situations, action is taken individually with domestic public policy by the two states; goals are not shared. Confrontation of the issue itself does not signify intention of collective action.

The next level of cooperation intensity is that of ad hoc joint action. For example, two states may be engaged in cleaning a shared river. Decision-makers of State A may negotiate to decrease pollutant levels, while those of State B may put forward plans to increase tourism in the basin, to which an improved ecological state of the river would be useful. This represents weak cooperation intention because their goals differ. There is joint action to clean up the river but it is taken up in an ad hoc manner.

Common goal formation is the third cooperation intensity on the TWINS matrix. This reflects the situation where actors share a goal, but do not converge on the type of joint action needed to achieve it. An example would be where two states decide to facilitate transportation along a shared river. Decision-makers of State X may argue that more infrastructure is needed, rapids cleared and riverbeds widened for ease of transport. However, decision-makers of State Y may insist on regulating traffic without the construction of new infrastructure but with better monitoring systems in place. So, despite there being consensus on the goal of enhanced river transportation, different actions are taken by the two states to achieve it.

Common norm formation is when there are common goals and norms, and agreed joint action. As the case studies later show, this level of cooperation intensity is often represented by an interaction based on norms institutionalized in treaties and agreements on water resources management. Common norms guide joint action to achieve common goals.

At the highest level of cooperation intensity, complete collective identity is formed. States do not differentiate between their domestic interests and their collective international interest, though complete collective identity formation is rare (Wendt 1999). This makes the highest level of cooperative interaction unusual.

The scales of conflict and cooperation intensities are summarized below (Figure 3.3). As mentioned, high intensities of coexisting conflict and

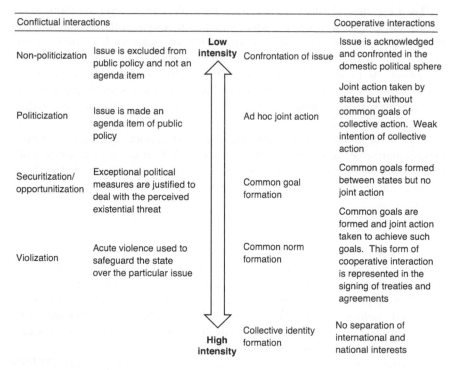

Figure 3.3 Summary of conflict and cooperation intensities.

cooperation (that is, war and unification respectively) are empirically unlikely in transboundary water interaction, because of the difficulties associated with sustaining such a unique condition.

Material capability and discursive power

As the previous two sections showed, scales of conflict and cooperation intensity qualitatively evaluate how transboundary water interactions form and change inter-state relationships. It should be emphasized that speech acts (assertive, directive and commissive) are *not* normative, and careful analysis is required to understand their impact. For example, a commissive speech act does not necessarily mean an improvement in the quality of the relationship. This is because power asymmetry may determine the extent to which cooperation is achieved, how, when and why (Zeitoun and Mirumachi 2008; Zeitoun *et al.* 2011). The analyst must question how the elite decision-makers frame the issue of water allocation for what purposes. Power analysis is crucial to understanding why cooperation may *not* provide equitable outcomes for all the basin states, and what the effects of 'cooperative mechanisms' that exclude certain issues or basin states may imply. Such analysis can also demonstrate that 'conflict' may bring new issues to the negotiation table that guide further transboundary water interaction.

To understand power asymmetry, the study utilizes the Framework of Hydro-Hegemony (FHH) (Zeitoun and Warner 2006) (see also Chapter 2). FHH identified three key factors (or 'pillars' according to Zeitoun and Warner 2006) that make hydro-hegemons to command control over the shared waters: riparian position; power and exploitation potential. The relative power of hydro-hegemons enables greater control of water resources compared to the other states. In the TWINS framework, the FHH factors, riparian position and exploitation potential are combined to represent material capability, and power is more specifically understood as discursive power. These two explanatory factors combined help to analyse the ways in which access and control over water resources are contested, negotiated and agreed.

Material capability is exemplified by the technological know-how and potential to abstract water resources and modify river flow. The extent of hydraulic infrastructure construction in the basin; the number of master plans for development; the availability, and level, of professional knowledge and human resources; and the degree of access to funding are useful indications of material capability. In this regard, material capability is closely linked to the maturity of the basin's hydraulic mission. The hydraulic mission represents a water resources management paradigm in which centralized efforts to engineer the river are characteristic (Reisner 1993; Allan 2001, 2003; Turton and Meissner 2002; Swyngedouw 2004; Wester 2008; Molle *et al.* 2009). Infrastructure is used to increase the volume of water stored within the basin and to manage it for multiple purposes, ranging from irrigation and flood control to hydropower development (Molle *et al.* 2009: 333). As centrally planned enterprises, these costly projects often achieve a symbolic effect of progress and industrial modernity (Wester 2008). Basin states may exercise their material capability in a way that acutely focuses on absolute quantitative allocation. The implication could be that issues of payment for ecosystem services, for example, are ignored and not discussed within this exploitative paradigm of water resources management. Advantages of the riparian position can also help bolster material capability. For example, the proximity to water resources could be used to effectively limit access by other, more distant basin states. Geographical characteristics of the basin state are examined not from the perspective of environmental determinism, but rather with the aim of understanding the way in which actors can capitalize on their position, and the extent to which it matters.

The TWINS framework focuses on discursive power in order to determine the performative effects (Guzzini 2005: 515) that change levels of conflict and cooperation intensities. Discursive power is observed through the ways in which persuasion, deliberation and consent occur over water allocation and utilization. This study focuses on an expression of soft power; that is, the discursive power exercised by actors to frame water issues to suit their own interests. This form of power, in contrast to the use of violence or hard power, is often used to justify outcomes of water resource allocation and development (Zeitoun *et al.* 2011). If there is agreement to cooperative initiatives to manage water, the outcomes become less prone to being challenged, regardless of project sustainability or equity. Speech acts that 'help constitute or change existing discourse coalitions and/or change an existing discourse, thereby reconfiguring existing relations of power' are useful indicators of the ways in which discursive power may be used (Stritzel 2007: 370). Soft power refers to non-material power highlighted in FHH: bargaining power to influence agendas; and ideational power to bring about compliance without contestation (Zeitoun and Warner 2006) (see Chapter 2). Soft power may be used for distributive ends, where quick political agreements are reached by one state deferring to a more powerful state; the outcome being superficial solutions to shared water resources management, which the non-hydro-hegemon state does not entirely buy into (Zeitoun *et al.* 2011). Soft power may also be exercised to bring about supposedly consensual outcomes where the hegemonic order that reproduces inequalities is not even discussed or questioned (ibid.). Conflict, in the form of diplomatic disagreement, can represent a challenge to the discursive power imposed by the hydro-hegemon.

This study proposes that basin states use public ideas and norms to promote, or demote, water resources development on the bilateral and multilateral political agenda, thereby leading to both conflictual and cooperative interaction. While it is argued that ideational power is the most effective (Zeitoun and Warner 2006), identifying the cause and effect of such power is difficult. However, through discourse analysis it is possible to examine how ideas and norms are operationalized, and translated into water resources management practice. Securitization represents one form of discursive power being put into practice. In addition, knowledge construction reflects the way in which discourses can formulate facts, regardless of their accuracy and scientific rigour. In other words, these 'facts' are not questioned and taken for granted, as a form of wisdom and popular belief (Zeitoun and Warner 2006: 448). Sanctioned discourse is another way of ensuring compliance by only allowing deliberation on issues which have been mooted by the hydro-hegemon, thus eliminating the political space to discuss alternatives (ibid.; Allan 2003).

Multilayer analysis of transboundary water interactions

It was explained earlier that the focus on elite decision-makers enables a better insight into the 'black box' of state decisions. Another advantage of focusing on this set of actors is the incorporation of spatial scale into transboundary water interaction. State elites are influenced by domestic factors when decision-making, including pressure from their constituents, budgetary constraints and personal gain. In addition, external factors, such as donor initiatives, may come into play. Meanwhile, decisions at the international basin level influence water use and allocation at the national level. The focus on elite decision-making enables better investigation into how the 'domain of hydropolitics' and 'domain of national policy-making' are linked, as pointed out by Mollinga (2008a, 2008b). While it is not possible to completely uncover the motivations of actors, especially those of a personal nature, unpacking the state by identifying these elite decision-makers provides an analysis of transboundary water politics that is more sensitive to the scalar implications of decision-making over shared waters.

In order to understand the role of spatial scales in transboundary water interactions, this study is further informed by contributions made in the field of political geography. Mustafa (2007) argued that hydropolitics focusing at the basin level is much too crude. Drawing on the Indus River basin, he emphasized that the analysis of power relations at the local level gives a better understanding of how water impinges upon the security of basin states. The concept of water and conflict geographies by Harris (2002) usefully underscores the contemporary–historical, vertical–horizontal and political–ecological linkages of water use and allocation. This concept does position conflict separate from cooperation, unlike the arguments I have laid out above. However, her insights on the Southeastern Anatolia Project or the GAP project by the Turkish government in the Tigris-Euphrates River basin is useful. Harris (2002: 745) argued that the research question should not be to ask '"will the GAP project result in conflict?"' Rather, she pointed out the question should be, '"in what ways is GAP water development related to historical, on-going, or future modes and sites of conflict, and what is the importance of these relationships for social, political, and economic processes at different temporal and spatial scales?"' (ibid.). Her study emphasizes that scale-sensitive analysis helps shift attention from the outcome to the *process* of conflict and cooperation.

That is not to say that hydropolitics, informed by IR, is completely unaware of spatial scale in its analysis.[8] Weinthal's (2003) work on Central Asia shows how domestic drivers contribute to the way in which water is used as a form of state-making. Waterbury (2002) sought explanations in domestic politics for the success, or failure, of multilateral cooperation in the Nile River basin. Directly relevant to the TWINS framework on coexisting conflict and cooperation is work by Warner (2008). His study pointed out that the state practice of coercion and gaining consent, or hegemony, transcends multiple spatial scales. He showed how national governments practise coercion and gaining consent in the domestic sphere, in order to build support for the political regime among different, fragmented political groups. At the same time, hydro-hegemony is directed at downstream states in an attempt to legitimize specific river basin development projects as providing a regional public good. The Turkish state actively maintains regional hegemony by putting forward projects aimed at peace-building. None the less, the Turkish government cannot be deaf to global discourses about good practices of water resources management and dam-building, and the global networks of NGOs and donor communities that frame the GAP project in terms of human rights and ethnic minority issues. Warner's (2008) work reminds us how power is practised across scales, so as to maintain and to challenge access and control over water resources.

The TWINS framework, it is argued, is an effective way of shedding light on politics about and over spatial scales that exposes socio-economic and political contexts. Analysis of transboundary water interactions necessarily entails examination of the broader, socio-economic and political context of water issues (Zeitoun and Mirumachi 2008; Zeitoun *et al.* 2011). The TWINS approach to transboundary water politics explains why certain scales are privileged over others in order to manage and govern shared waters by different actors. Again, the construction of scale is another key element that may be gleaned from political geography. Of note is work by Harris and Alatout (2010) who argued that state actors appropriate scales to water resources management and governance to promote political agendas on state-building and territorial concerns. Their work is useful because it emphasizes that discourse and practice are constantly at work, with state actors attempting to stabilize and challenge scales at which water issues are 'best' managed for their political ends.

Sneddon and Fox (2006) go further to extending analysis beyond state actors and offer a research agenda for what they term 'critical hydropolitics'. They critiqued work on the hydropolitics of shared rivers as narrow and state-centric, and ignorant of the scalar implications of interventions. They argued insightfully that in fact, cooperation among basin states in the form of institutionalized arrangements usually overlooks socio-economic, political and ecological implications at the local level. This claim identifies the range of discursive frames of cooperation over natural resources, advocated by different actors. For state actors, 'cooperation is perceived as the basis for proceeding with the *development* of water resources encompassed by basins' (ibid.: 182; emphasis in original). On the other hand, local communities may frame the river as a resource for the sustenance of livelihoods, necessitating a spatially different scale perspective of the river and its resources than that of the decision-maker in the government (ibid.).

Intertwined with the issue of construction of spatial scales is the role geographical imagination plays in informing transboundary water interactions. Geographic imaginations of the river and the river basin can project ideas of water scarcity, abundance, threats and opportunities. Dalby (2009) argued that geopolitical imaginations of fear of the 'other' threatening resource scarcity can also facilitate a certain geopolitical order. This geopolitical order is another mechanism through which power relations may be observed. Considering critical geopolitical perspectives adds to analysis on how decision-makers justify and legitimize the 'best' scale for transboundary water resources management. In other words, the historical and political context of constructions of threat, something overlooked in the Copenhagen school of securitization theory (Mason and Zeitoun 2013), helps explain the solutions and measures taken up in elite decision-making.

Conclusions

This chapter has explained the conceptual framework of TWINS, which examines coexisting conflict and cooperation within international transboundary water interaction. The four-by-five TWINS matrix of conflict and cooperation intensity levels is a way to avoid over-simplified characterizations of basins in *either* conflict *or* cooperation. By focusing on speech acts, analysts using the TWINS matrix can trace the ways in which relationships between basin states can change over time. Basin asymmetries in material capability, and the exercising of discursive power, help explain these changes in conflict and cooperation intensity. The TWINS framework also enables some advances in the way agency is considered within hydropolitics. Elite decision-makers of the hydrocracy and those associated with it are key actors shaping national interests and positions within international negotiations. This original framework can also accommodate consideration of multiple spatial scales in which politics occurs. What is projected at the international fora is mediated by, and through, domestic politics – where power struggles between sectors and stakeholders occur. The next three chapters apply TWINS to case studies from the Ganges, Orange–Senqu and Mekong river basins.

Notes

1 Acknowledgement is given to Allan (2007) for this example of marriage, which came up in a London Water Research Group annual workshop: 'those of you who have a close relationship in a marriage or similar relationship know that conflict and cooperation go on at the same time. It's *normal*' (emphasis in original).
2 See Mirumachi (2007a, 2007b) for an earlier version of the framework; also Mirumachi and Allan (2007).
3 See an application of this particular TWINS interpretation on issues of desalination between Israel and Jordan in Aviram *et al.* (2014).
4 The recipient of the speech act is termed as 'audience' within speech act theory.
5 The databases may be found online: www.transboundarywaters.orst.edu/database/interwatereventdata.html http://worldwater.org/water-conflict/.
6 See Mirumachi (2007a, 2007b), Mirumachi and Allan (2007) for earlier versions of conflict and cooperation intensity scales.
7 Tuomela's (2000) points about intention, joint action and dependency are congruent with Wendt's (1999) constructivist notions of collective identify. Wendt (1999: 343) argued that interdependence, common fate, homogeneity and self-restraint facilitate collective identity formation.
8 See also a collection of essays on the theme of global environmental governance within IR, and the achievements of, and challenges to, incorporating scale (O'Neill in Trudeau *et al.* 2013).

4 Securing and securitizing cooperation in the Ganges River basin

Introduction

The Ganges River is deemed as being endangered as a result of the over-abstraction of water resources, making it one of the top ten rivers in the world to be 'at risk' (Wong *et al.* 2007). It is particularly exploited compared to other rivers making up the larger Ganges–Brahmaputra–Meghna (GBM) River basin, shared by China, Bhutan, Nepal, Myanmar, India and Bangladesh (Babel and Wahid 2008: 15) (see Figure 4.1). The freshwater flow of the Ganges River benefits from the snow and glaciers of the Himalayan Mountains, often described as Asia's water towers. As a basin that is comparatively abundant in water resources, the river has provided opportunities for irrigation expansion and agricultural development that has led to over-abstraction. This problem is compounded by the actual and potential impacts from hydrological changes taking place as a result of climate change

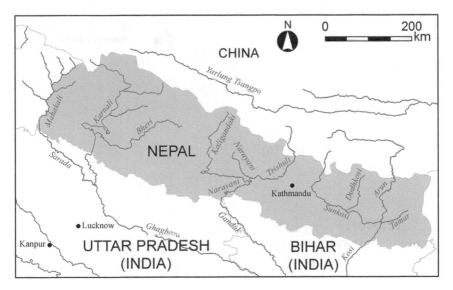

Figure 4.1 Map of the Ganges River basin.

(Collins *et al.* 2013; Immerzeel *et al.* 2013; Jeuland *et al.* 2013). Considering that the GBM basin is home to some 633 million people (FAO 2012: 113), the implications of hydrological change and water availability could be far-reaching, for both the ecosystems and the livelihoods that depend upon it.

With this in mind, the Nepal–India relationship provides an entry point for analysis of conflict and cooperation over shared waters in one of the most populous river basins in the world. Of the 1,087,300 km² basin, India occupies the largest area with 860,000 km², followed by Nepal with 147,500 km² (FAO 2012: 111). The mean annual runoff of the Ganges River is estimated to be 415,000 million m³ at the Farakka Barrage downstream in Bangladesh; though with monsoon rains the river flow is highly variable throughout the year (Mizra 2004: 16). The tributaries in Nepal contribute to this annual flow with 210,200 million m³ from four major tributary rivers and other basins in the south of its territory (FAO 2012: 366). The tributary rivers – the Karnali, the Kosi and the Gandak – play an important role, particularly in contributing to dry season flow, in addition to the Mahakali River forming a boundary between India and Nepal (Adhikary *et al.* 2000). The wet season between June and September often triggers flooding and the common phrase 'Bihar's sorrow' reflects the severity of flood hazards for local communities along the Kosi River. Transboundary water interaction between Nepal and India spans more than half a century, particularly over multi-purpose projects to control the river and to exploit water resources. This chapter analyses the river engineering projects and water utilization plans that have led to high levels of abstraction.

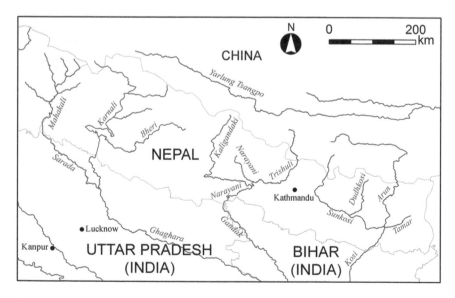

Figure 4.2 Map of the Ganges tributaries.

Insight gained from this transboundary water relationship informs exam-ination of water governance of the larger GBM River basin. What kinds of challenges do rivers 'at risk' like the Ganges River bring about for larger scale governance? Water use is an increasingly debated issue among the six basin states of the GBM basin. Transboundary water interactions between Bangla-desh and India have contested water use, most notably represented in the negotiation of the Ganges Treaty. China has had little formal transboundary water interaction with other basin states, although there are now reports of increased interest in dam development, which would more extensively exploit the resources of the Asian water towers (Pomeranz 2013). Both Bhutan and Myanmar are also increasingly becoming heavily vested in hydro-power development, taking advantage of the available water resources. I argue that the constraints to multilateral water governance may in part be attributed to a distinct path dependency of water resources management within the region, underpinned by power asymmetries of basin states. As the following sections will show, a critical analysis of 'cooperative' water arrange-ments put in place reveals inequalities of water allocation and missed oppor-tunities for addressing the risks of over-abstraction.

Establishing institutions for bilateral water resources development

Developing hydraulic infrastructure to control water flow on the Ganges trib-utaries characterizes the initial stages of riparian interaction between Nepal and India. This development was guided by agreements pertaining to specific bilateral projects. Although agreement was reached between Nepal and British India in 1920 regarding the Sarada River, the Kosi Agreement (1954) and the Gandak Agreement (1959) marked the beginning of formally institu-tionalizing transboundary water use by the two independent states.[1] As a major tributary of the Ganges River system, the Kosi River was a concern for the governments of both Nepal and India, particularly because of its tendency to cause floods (see Figure 4.2). The Indian government considered measures to improve the situation in the state of Bihar, but severe flooding in 1954 triggered plans to develop a barrage to regulate flow in Nepali territory (Verghese 1999: 339–340). Consequently, as a means of reducing flooding, the *Agreement between His Majesty's Government of Nepal and the Government of India concerning the Kosi Project* (hereinafter the Kosi Agreement) was signed by the two states.

This project was also to provide ways to use the river for irrigation and energy production. It was also anticipated that the barrage would help reduce soil erosion in Nepal (Kosi Agreement 1954: Preamble). Representative of a commissive speech act, the Agreement established common norms of water use. The Kosi Agreement also established rules for project implementation and operation, allocating costs and rights to water resources utilization. The multi-purpose nature of the barrage reflected the hydrocracies' view that the river was a source of environmental threat (i.e. from flooding and erosion) but also an opportunity to augment the economic value of water through irrigation and hydropower. Thus, common norms governing water use in this case attempt to maximize economic benefits and reduce environmental threats (TWINS Sequence 1 in Figure 4.3).

This view of the hydrocracies also informed the development of the Gandak River. In 1957, a preliminary agreement was signed between the two governments to jointly develop a multi-purpose barrage project on a section of the Gandak River that forms the international boundary. The project was designed to take advantage of regulated water flow by constructing canals in both Nepali and Indian territories: the Western Nepal Canal and the Indian Don Branch Canal respectively.[2] This agreement was formalized in the *Agreement between His Majesty's Government of Nepal and the Government of India on the Gandak Irrigation and Power Project* (hereinafter the Gandak Agreement), signed in 1959. This bilateral agreement clearly expressed that the development of the river was for 'common interests' in using the shared waters (Gandak Agreement 1959: Preamble) (TWINS Sequence 2 and 2' in Figure 4.3).

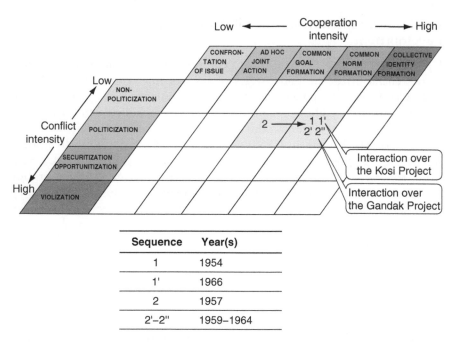

Sequence	Year(s)
1	1954
1'	1966
2	1957
2'–2"	1959–1964

Figure 4.3 TWINS matrix of Nepal–India relationship: Sequences 1–2.

Note
Numbers in the matrix represent sequence of significant speech acts. The position of sequence numbers within a cell is for presentation purposes only and it does not represent degrees within a certain conflict or cooperation intensity level. For example, even though Sequence 1 is placed at the left side of the cell, it is the same level of conflict and cooperation intensities as Sequence 2'.

However, sharing the benefits from water use was highly politicized for both projects. In Nepal, there were concerns about lack of irrigation benefits and unclear provision with regard to cost sharing (Untawale 1974: 719). The Kosi project was an Indian-owned project. The Kosi Agreement specified that the land and water rights of the project site were entitled to the Indian government. Moreover, water flow and hydropower generation were controlled by the Indian government at the barrage site near Hanuman Nagar in Nepal. Hydropower was available to Nepal at a fee for a maximum of 50 per cent of the capacity generated (Kosi Agreement 1954: Article 4). The Agreement specified that the construction material obtained from within Nepal would be compensated by India (ibid.: Article 6).

The international agreement had a major impact on the domestic politics of Nepal. There was public opinion that the Kosi Agreement was unfair and that the government was 'bartering away Nepal's future' (Jha 1973: 43). The agreement was so controversial that it became politicized, with the Nepali monarch, King Mahendra, making a directive speech act to the Indian government for amendments. This speech act triggered bilateral meetings to be initiated and the Kosi Agreement was finally amended in 1966 (Dhungel 2009: 26) (TWINS Sequence 1' in Figure 4.3). The new agreement addressed Nepali concerns about sovereignty and territorial loss. Arrangements were made so that instead of an indefinite period of land lease to India, a lease of 199 years was stipulated (Amended Kosi Agreement 1966: Article 5). In addition, the Nepali government gained an increased degree of freedom with regard to the control of water resources, including the tributaries of the Kosi River, thereby providing more opportunities for irrigation (ibid.: Article 4).

The politicized nature of water resources development is further reflected in the clauses of the Gandak Agreement, as it incorporated lessons learned from the controversy surrounding the Kosi Agreement. The Gandak Agreement specified water allocation, entitlement to hydropower benefits and a guarantee of sovereignty for Nepal (Salman and Uprety 2002: 89). It also stipulated that the Indian government was responsible for the construction of the two canals in Nepal, each with a minimum flow of 20 cusecs. This was in addition to the development of a 15,000 KW power station in Nepal using water flowing through the Indian Western Main Canal, and the distribution of generated power through a power grid (Gandak Agreement 1959: Articles 7, 8). None the less, issues of inequality of benefits were deemed problematic, with strong opinion within the Nepali government and among the public that the nation was being exploited (Jha 1973; Untawale 1974). The Gandak Agreement specified water allocation quantities for Nepal but there were no figures that indicated water allocation for India, raising doubts about the extent of benefits (Dhungel 2009: 23). Moreover, Nepal's water use on the Gandak River was ultimately restricted because it had to ensure water flow of a predetermined monthly volume to the two project canals (see Gandak Agreement 1959: Annex).

This public perception and the negotiations point to the difficulty of establishing quantitative water allocation and equity of benefit sharing. In parallel to the negotiations concerning amendments to the Kosi Agreement, changes were made to the Gandak Agreement in 1964 (TWINS Sequence 2" in Figure 4.3). The amended agreement gave increased freedom for Nepal to use water from the Gandak River system within its territory. Water use was recognized without restriction albeit trans-valley water use during the dry season months that would require separate agreements (Amended Gandak Agreement 1964: Article 9). Furthermore, Nepal would have control over the regulator for the Indian Don Branch Canal feeding the Gandak waters, in order to ensure sufficient water flow in the Eastern Nepal Canal which extends from the Indian canal (ibid.: Article 7).

The amendments of both the Kosi and Gandak agreements represent shared intentions to develop the river but unclear mechanisms for measuring equity. For example, despite the amendment, the Kosi Project presents a stark contrast to irrigation opportunities. On one hand, for India, the Kosi Eastern Canal, which is entirely in its territory, provides irrigation for an area of 612,500 ha. The Kosi Western Canal provides a further 356,610 ha of irrigation capacity. On the other hand, for Nepal, the 35 km of the Kosi Western Canal constructed in its territory enables water to 11,300 ha (Salman and Uprety 2002: 69–70). Moreover, the construction, which was the responsibility of India, was delayed (Dhungel 2009: 19; see also Documents 694, 691 in Bhasin 2005). This lack of clarity is glossed over by assertive speech acts by the Indian hydrocracy, as evidenced in the following remark by the Indian Irrigation Secretary, C.C. Patel, recorded at a bilateral meeting on water resources development in 1979:

> [R]ivers common to Nepal and India, which are God's gifts, should be developed for the mutual benefit. These rivers have great potential for destruction as well as for the benefit. By mutual cooperation and action, the destructive aspects could be eliminated and the benefits derived by way of irrigation, hydro-power, flood control etc. [...] Shri Patel mentioned that the Kosi is known as the River of Sorrow for both Nepal and India and unless a cooperative approach is adopted for harnessing Kosi waters, the recurring damages cannot be avoided.
>
> (Minutes, December 1979 (Document 699) in Bhasin 2005: 1371–1372)

This speech act begins to normalize bilateral cooperation that justified joint river basin development as a way to avert hazards, and has the effect of underplaying the politicized nature of water allocation. Hydro-hegemony of India is apparent, with the level of control over the river flows and opportunities to develop the river basin. This normalization of bilateral cooperation becomes even more evident in further negotiations over larger projects, explained below.

Large-scale projects for river basin development

The Kosi and Gandak projects established the first phase of bilateral negotiations over shared waters in the development of transboundary water relations between Nepal and India. A second phase followed suit, this time focusing on much larger multi-purpose projects during the 1960s to 1980s. Of particular note are the Karnali Project, the Pancheshwar Hydro Electric Project and the Kosi High Dam Project. The Karnali and the Pancheshwar Hydro Electric projects had a major hydropower component, while the Kosi High Dam Project aimed to reduce flooding.

The Karnali Project was proposed in the early 1960s by the Indian government, with the goal of generating hydropower and storage for irrigation water on the Karnali River in Nepal. Because this was a massive project with a potential 10,800 MW capacity, it was highly prioritized in bilateral political discussions throughout the 1960s and 1970s (Dhungel 2009: 31). The Pancheshwar Hydro Electric Project was proposed by the Indian government in 1977 as an important political agenda to reaffirm bilateral cooperation (Joint Press Statement in Dhungel and Pun 2009: 373). This 2,000 MW project on the Mahakali River (named as Sarada in Indian territory) would provide irrigation and flood control. Bilateral negotiations focused on how water resources and the hydropower/flood control benefits of the project could be shared. The Indian government again put forward plans, this time the Kosi High Dam Project in the early 1980s. The initial plan was to construct a high dam in Barakshetra, Nepal, which could help reduce flooding.

The negotiations over these projects signified commissive speech acts by both the Indian and Nepali hydrocracies and elite decision-makers, despite it being a prolonged process (TWINS Sequence 3 in Figure 4.4). Shared tributaries were conceived as ideal geographies for exploiting the abundant water resources, although this time through bilateral projects that were much larger in scope and more expensive than the Kosi and Gandak projects. However, these commissive speech acts also coexist with tensions regarding the ownership of the project. For example, even though the Karnali Project was suggested by India, the Nepali decision-makers counter-proposed with revised plans for a domestic project, from which India could purchase power. In this way the Nepali government could secure ownership of the project, something they failed to do with the Kosi and Gandak projects (Pokharel 1991: 81). During the negotiations over the Pancheshwar Hydro Electric Project, the Nepali delegation insisted that the benefits of hydropower, irrigation and flood control should be divided equally between Nepal and India. This was argued based on the fact that the project would be implemented on a shared, common river. India's hydro-hegemony may be seen in the way it can table options for development, which then becomes the basis for negotiations for Nepal. There are occasions where Nepali decision-makers attempted to challenge some proposals. For example, the Indian hydrocracy made an assertive speech act indicating that the Kosi High Dam was a necessary solution to

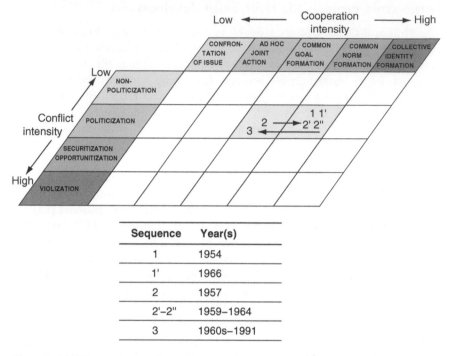

Sequence	Year(s)
1	1954
1'	1966
2	1957
2'–2"	1959–1964
3	1960s–1991

Figure 4.4 TWINS matrix of Nepal–India relationship: Sequence 3.

problems of excessive sediment and recurring flooding (Minutes, April 1983 (Document 707) in Bhasin 2005). In response, the Nepali hydrocracy argued that it should be constructed as a multi-purpose dam with hydropower and irrigation benefits, effectively rejecting the idea that the high dam could eliminate flood risks (Record, September 1984 (Document 716), Minutes, December 1987 (Document 718) in Bhasin 2005). Counter-proposals such as these reflect the strategic moves by Nepali decision-makers to maximize benefits from joint projects in situations of power asymmetry.

Securitizing the Tanakpur Barrage issue

In addition to the Pancheshwar Hydro Electric Project, the Indian government planned to further develop the Mahakali/Sarada River with the Tanakpur Barrage project. The Tanakpur Barrage was to sustain irrigation activities further downstream in Indian territory, as the Sarada Barrage became obsolete.[3] The Sarada Barrage facilitated agricultural development of Uttar Pradesh, and new infrastructure was needed to continue regulating water flow. However, the Nepali hydrocracy was concerned about the proposed Tanakpur Barrage for several reasons. First, there was an issue of resolving problems stemming from historical arrangements of the existing Sarada

Barrage before the Tanakpur Barrage could be constructed as a replacement for it. The Sarada Barrage was built in what was originally Nepal acquired by colonial India as part of a land exchange. The Sarada Treaty in 1920 granted the Nepali government access to a minimum of 400 cusec and a maximum of 1,000 cusec of water from the Sarada Canal. From the Nepali hydrocracy's perspective, if the Tanakpur Barrage was to regulate water further upstream of the Sarada Barrage there needed to be clarification about the equity of water allocation from the Mahakali River.

Second, there were complexities surrounding the design of the project. The Indian hydrocracy found that the project had to be designed so that one component was built in Nepali territory. Essentially, the left afflux bund, or the wall-like structure of the barrage, had to be sited on higher ground east of the Nepali border. As was experienced during the Kosi Project, issues of land and territory were bound up in the discussion about water use. Potential inundation and erosion of land were other issues, which exacerbated concerns about loss of territory, and control over sovereignty. These concerns were reflected in directive speech acts by the Nepali hydrocracy between 1983 and 1987. These speech acts attempted to frame the Tanakpur Barrage project as an *international* project on Nepali soil, and that the hydropower component should not negatively impact upon Nepali irrigation water allocation (Documents 707, 716, 718 in Bhasin 2005).

To respond to these concerns, the Indian hydrocracy initially agreed to install a head regulator for Nepal so that it has access to 1,000 cusec of water, and to share 25 MW of hydropower. However, the Indian hydrocracy also made a directive speech act by proceeding with the construction of the project in its own territory by the late 1980s, despite Nepali reservations (Letter, May 1991 (Document 727) in Bhasin 2005). Furthermore, the Indian government made a securitizing move in which the project was taken out of the realm of normal politics into one of 'emergency politics' (Roe 2006: 426) that justified urgent action. The securitizing speech act by the Indian Prime Minister, Chandra Shekhar, had the effect of increasing the conflict intensity of transboundary water interactions:

> While the unresolved issue could be formally taken up in the [Nepal–India] Joint Commission meeting [on general water resources issues], in view of the approaching monsoon, the work of the left afflux bund has to be completed at the earliest. The areas at the border on the left side of the river at Tanakpur are subject to inundation and erosion, and tying the left afflux bund with high banks in the Nepalese territory, as proposed by us, will bring a permanent solution.
>
> (Letter, May 1991 (Document 727) in Bhasin 2005: 1554–1555)

With this speech act, the construction of the left afflux bund into Nepali territory was presented as a *necessary* measure against threats to both states. Moreover, this measure was deemed a suitable 'permanent solution', thus avoiding the need for any temporary action. Securitizing the project in this way framed it as the *only* option available against the oncoming ravages of nature. This speech act meets the conditions of securitization according to Buzan *et al.* (1998: 32–33), with the safety of the two states passing the 'point of no return' unless action is taken with the only viable measure (TWINS Sequence 4 in Figure 4.5).

As pointed out in Chapter 3, threats to securitize do not have to be real, and this case presented by the Indian government proved very persuasive. Faced with only one, necessary and urgent option, the Nepali government granted permission for the construction of the left afflux bund in December 1991. A document by the Ministry of Foreign Affairs to the Indian Embassy in Nepal communicated agreement to the bund, which would be 577 m long. It was noted that 2.9 ha of Nepali land would be made available for the construction (Note, December 1991 (Document 730) in Bhasin 2005). This understanding was confirmed in a Memorandum of Understanding (MOU) signed by the two countries. This securitizing practice established the rules of water use and benefit sharing. Specifically, Nepal was allocated 1,000 cusec of water and 10 MW of hydropower. Securitization had the effect of 'closing down' discursive space for any alternatives of ways to share benefits. Moreover, it was made material by the Indian government constructing the concrete

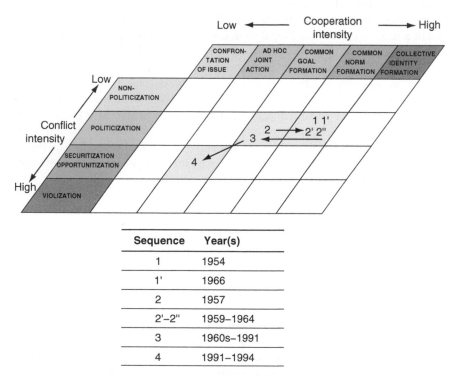

Sequence	Year(s)
1	1954
1'	1966
2	1957
2'–2"	1959–1964
3	1960s–1991
4	1991–1994

Figure 4.5 TWINS matrix of Nepal–India relationship: Sequence 4.

left afflux bund. The Tanakpur Barrage is said to have been operational in 1990/1991 and the left afflux bund component in Nepali was finished before the monsoon season in 1992 (Dhungel 2009: 42–43; see also Salman and Uprety 2002: 102–103; Press release, February 1993 (Document 742) in Bhasin 2005).

The bilateral interaction over the Tanakpur Barrage may be described as one of coexisting securitization and ad hoc joint action. The securitizing practice put into place clear goals for river basin development but there was weak intention for collective action (see Chapter 3). This was owing to the major controversy over the project at the sub-national level in Nepal, which brought about fragmented decision-making. The MOU was initially a useful tool for the Nepali elite decision-makers who negotiated the deal because it did not recognize the Tanakpur Barrage arrangement as an international agreement with India. That being so, it did not require parliamentary ratification, as is stipulated by the Nepali Constitution, for all international agreements. None the less, there was strong objection to this decision from politicians and the public, considered to confer territorial rights with little compensation (Dixit 1997: 159–160). The Prime Minister, G.P. Koirala, was challenged about this decision by protest rallies and strikes, and pressure mounted on the government.[4]

The controversy was such that the Supreme Court of Nepal adjudicated on the legal status of the Tanakpur arrangement. The ruling in 1992 determined that the MOU was, in fact, a bilateral agreement requiring majority parliamentary approval. Despite this verdict, political controversy ensued over the degree of consensus required within Parliament concerning this matter. If the Tanakpur Barrage issue was a 'pervasive, simple and long-term' issue – as specified in Article 126, Clause 2 of the Constitution – it would require a two-thirds majority; otherwise a simple majority would suffice. The extent of discontent over the project was apparent when a special commission, commonly known as the Baral Commission, was set up to provide clarity (Gyawali and Dixit 2000: 242–249). The Nepali parliament had yet to clarify the legal requirements surrounding the consensus of the Tanakpur barrage, much less agree on it. This confusion over the arrangement reflects the weak levels of intention over the Tanakpur Barrage.

The contention around the legal status of the Tanakpur Barrage project arrangement makes the construction of the afflux bund by the Indian hydrocracy all the more significant. Warner (2004a, 2004b, 2011) identified that 'emergency politics' are played out when vested interests are furthered by taking advantage of crisis, and not just in the case of averting it. This opportunitization enables an actor to pursue options usually not available. I argue that the unilateral construction of the project in Nepali territory by India under these circumstances of crisis in Nepal is such example. Institutionalizing water use on the Mahakali River was achieved in a way that normal conditions of bilateral negotiation would not allow. The Tanakpur Barrage was always intended to benefit irrigation downstream, especially as the Sarada

Barrage had become a permanent feature of water resources development. However, this legal dispute within Nepali domestic politics posed a unique opportunity to seek benefits of irrigation under the terms the Indian hydrocracy set out in the MOU and ensure hydro-hegemonic control over the project.

'Cooperation' as a result of securitization

The water development projects since the 1950s demonstrate how the Nepali hydrocracy had difficulty in establishing water control. This institution, supported by successive governments, had little power to fundamentally change the status quo of water allocation. The Nepali government that established the MOU lost power as a result of the internal unrest, and a succession of short-lived governments provided little opportunity for rescinding the initial agreement of 1991. This meant the Nepali government did not challenge the institutionalization of the costs and benefits. However, it sought to gain better bargaining power by proposing the Integrated Development of the Mahakali River. This outlined a framework agreement of existing and proposed projects on the Ganges tributaries. With a new arrangement that included several projects, the intention of the Nepali government was to secure increased water allocation and hydropower capacity from the Tanakpur Barrage, and from the proposed Pancheshwar Project (Gyawali and Dixit 2000: 251–252). The two governments conducted bilateral talks on this proposal during 1995 (TWINS Sequence 5 in Figure 4.6).

An agreement was reached the following year, enshrined in the *Treaty between His Majesty's Government of Nepal and the Government of India concerning the Integrated Development of the Mahakali River, including Sarada Barrage, Tanakpur Barrage and the Pancheshwar Project* (hereinafter the Mahakali Treaty). With the signing of the Mahakali Treaty, the two states had made commissive speech acts regarding water resources development not only for the Sarada, Tanakpur and Pancheshwar projects, but also for future projects on the Mahakali River. The Mahakali Treaty defined:

> equal entitlement in the utilization of the waters of the Mahakali River without prejudice to their respective existing consumptive uses of the waters of the Mahakali River.
>
> (Mahakali Treaty 1996: Article 3)

The Treaty also specified the benefits to Nepal from the Tanakpur Barrage, such as free hydropower of 70 GW hours. Water was allocated at 1,000 cusec during the wet season and 300 cusec during the dry season (Mahakali Treaty 1996: Article 2). In addition, Nepal was entitled to 1,000 cusec and 150 cusec during the wet and dry season respectively from the Sarada Barrage (ibid.: Article 1). With regard to the Pancheshwar Project, the hydropower stations on both the Indian and Nepali sides of the river give the two states equal ownership and equal share of the generated hydropower with costs shared proportionate to the benefits (ibid.: Article 3).

One of the significant aspects of the Mahakali Treaty was that it recognized 'equal partnership' between Nepal and India (ibid.: Preamble). In addition, the principles of the treaty acted as a template for the design and implementation of other projects on the Mahakali River (ibid.: Article 6). After the Mahakali Treaty was signed, the Pancheshwar Project was seen to symbolize cooperative water resources development between Nepal and India. Nepali and Indian politicians stated publicly that the Treaty amounted to definitive progress in bilateral water resources development. The Nepali Minister of Water Resources asserted that a 'landmark treaty' had been achieved (FT Energy Newsletters 1996: 11). Within Nepal, parliamentary approval was gained, unlike with the Tanakpur Agreement (though there was much political dispute between different parties and strong aversion to ratification) (Kumar 2004). The Indian Minister of Water Resources also confirmed the Treaty as 'a landmark in the long history of close ties between the two countries' (Statement, November 1996 (Document 772) in Bhasin 2005: 1677).

The treaty established common norms of 'equal partnership' for the achievement of the common goals of water resources development. Transboundary water interactions changed from high conflict–low cooperation interaction to medium conflict–high cooperation interaction (TWINS Sequence 6 in Figure 4.6). However, I argue that the Mahakali Treaty merely legitimized the securitizing move the Indian government had taken to consolidate its control over water resources, through the construction of the Tanakpur Barrage. The status quo of water allocation was not replaced by a more transparent arrangement that took into account existing water use from the various projects under the Mahakali Treaty; as the analysis showed, water utilization was highly skewed between the two states. Without a clear understanding of the extent of abstraction already being done for various domestic water development projects, it is not possible to determine the equity of benefits of water use, beyond what is agreed specifically at the project level. This is a significant point because it shows that medium conflict–high cooperation interaction can still involve issues of inequality over water allocation.

Cooperation as a result of securitization is also significant in that it served to simply draw out the negotiations over the specifics of the Pancheshwar Project. The Detailed Project Report (DPR) that was to set out the parameters of the project was delayed and effectively abandoned, despite an understanding that it would be completed within six months after the signing of the Mahakali Treaty (Letter of Exchange 1996). One particular fundamental issue marked the deliberation: the role of the hydropower component. The Nepali hydrocracy argued strongly that the hydropower component would supply peak energy (Letter, August 1997 (Document 783) in Bhasin 2005: 783). A peaking power plant with over 6,000 MW would benefit Nepal

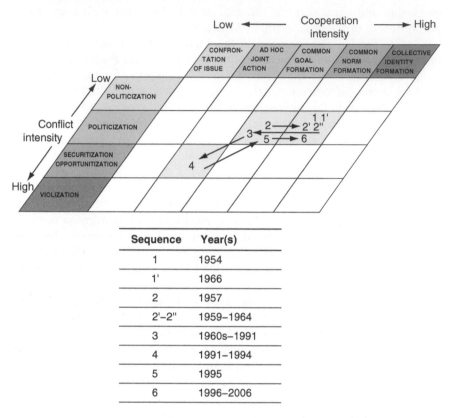

Figure 4.6 TWINS matrix of Nepal–India relationship: Sequences 5–6.

because peak energy has higher financial value than baseload power when trading energy with India (Marty 2001: 184). Indian decision-makers were reluctant to agree, preferring a hydropower component of 2,000 to 4,000 MW (Letter of Exchange 1996). It was only in late 1997 that a decision was reached on the construction of a peaking power plant (Summary Record, November 1997 (Document 786) in Bhasin 2005).

Common norms guiding water resources development on the Mahakali River were still in place: maximizing the efficiency of the water resources and sharing the benefits. A technical committee, the Joint Group of Experts, was established to determine the specifics of the Pancheshwar Project. However, the problems of ambiguity over existing consumptive water use could not be resolved. Nepal's rights to water during the wet and dry season were specified. In contrast, India's rights to water were not quantified in detail, except to stipulate responsibility for maintaining 350 cusec downstream of the Sarada Barrage (Mahakali Treaty 1996: Article 1). In 1997, the Indian government argued that the project also needed to ensure water flow in the

Lower Sarada Canal – 160 km downstream in India – as part of its existing consumptive use, despite not specifying the extent of water utilization. Negotiations stalled after Nepali decision-makers refused to accept use of water in the Lower Sarada Canal as indicating existing consumptive use (Kathmandu Post 1999). Despite the treaty being hailed as a new era of cooperation, transboundary water interactions following the Mahakali Treaty signing may be described as the 'Mahakali Impasse' (Gyawali and Dixit 2000). Article 12 of the Mahakali Treaty specifies that a review of the Treaty should be carried out every ten years. This meant that the two states had the opportunity in 2006 to clarify aspects of the Treaty, including the contentious issues around water allocation and benefit sharing. However, no official amendment was made. The TWINS analysis of this case study extends to this point in time, when an opportunity was missed to review the significance and utility of the bilateral treaty.

The political economy of transboundary water interaction

The TWINS analysis reveals a key feature of the transboundary water interaction between Nepal and India: small changes in cooperation intensity over time, which coexists with conflict. Reasons for transboundary water interaction exhibiting joint action or the development of common goals over water resources management may be found in the national hydraulic missions of the two states. For both Nepal and India, achieving the hydraulic mission required pursuing projects at national *and* transboundary scales. The means used by the two states to achieve their hydraulic mission reveal the national interests that led to contention over the interpretation of equity.

The Nepali government adopted a centralized approach to developing its water resources. This was achieved through national plans of infrastructure expansion, as specified in the National Five-Year Plan starting in 1956. These plans facilitated the construction of infrastructure needed for hydropower generation and irrigation expansion, aimed at development goals of 'meeting basic needs' and 'self-reliance' (Khadka 1988: 556). This explains the backdrop to the development of the Kosi and Gandak projects. The agreement amendments – examined in Sequences 1' and 2" in the TWINS matrix – reflect how the Nepali hydrocracy politicized the projects in order to advance its hydraulic mission and broader policy goals of economic development. Transboundary water interactions certainly led to some gains for Nepal. Insofar as the Kosi Project was concerned, the Chatara Canal Project was constructed and renovated, and the West Kosi Canal and accompanying pumping scheme came on line in the late 1970s. Although the project yielded much less irrigation benefit than originally planned, 98,000 ha of irrigation expansion was achieved. The Gandak Agreement provided the opportunity for the Nepali government to operate the Eastern Gandak Canal and to receive water from the Western Gandak Canal. As with the Kosi Project, the actual irrigation benefit was less than anticipated, but the two canals were able to irrigate an area of 39,000 ha (Dixit *et al.* 2004).

Hydropower development is a key feature of the Nepali hydraulic mission. In the 1980s, hydropower was identified as a means of generating foreign exchange revenue, making it an 'exportable commodity' (National Planning Commission 1984: 25). The desire to use water resources for state revenue resulted in the prioritization of multi-purpose hydropower projects. In the Sixth National Five-Year Plan (1980–1985) for Nepal, water resources development is described as the following:

> One of the basis [*sic*] policies governing electricity production will be to develop hydel power schemes, which can meet not only the long run power needs of the country but also create a surplus whose export can augment foreign exchange earning. Such projects will as far as possible be of the nature of multi-purpose projects, covering irrigation and other utilities.
>
> (National Planning Commission 1981: 93)

Nepal's strategic use of water resources may be identified in the establishment of its 1971 Power Exchange Agreement with India, and its request in 1988 for an increase in power exchange capacity to 25MW (Summary Record, December 1988 (Document 720) in Bhasin 2005). These policies help explain why the Karnali Project and, to a certain degree, the Pancheshwar Project were on the bilateral political agenda during the 1980s – as exemplified in Sequence 3 of the TWINS matrix. As Molle *et al.* (2009) argued, the hydraulic mission not only focuses on irrigation expansion but also on hydropower generation.

The Indian government also found river basin development crucial for economic development. In the early days of Indian independence, constructing concrete dams was a particularly symbolic venture for the state in its drive for economic development and modernization. As Khilnani (2003: 61) commented, 'India in the 1950s fell in love with the idea of concrete'. Agricultural development was also a national goal, with the aim of feeding an increasing population and to reduce poverty. To this end, water was an indispensable natural resource, as reflected in the First Five-Year Plan (1951–1956):

> The large land resources of India cannot be put to productive use without a simultaneous development and use of the water resources. In fact, an integrated development of the land and water resources of India is of fundamental importance to the country's economy.
>
> (Planning Commission 1952: ch. 26, para. 1)

During the First Five-Year Plan period, irrigation investment comprised 23 per cent of all planned expenditure, and the Ganges River basin was seen as the main target for enhanced irrigation (Planning Commission 1960; Narayanamoorthy 2005: 47). Coupled with the Green Revolution, the hydraulic mission facilitating water abstraction continued to improve agricultural productivity during the 1970s and 1980s. From the First Five-Year Plan to the Ninth Five-Year Plan (between 1950 and 2002), 308 major irrigation projects and 1,004 medium irrigation projects were proposed (Planning Commission 2002: 895). Although not all these projects were implemented or completed, these figures represent the grand scale of the Indian hydraulic mission. Post-independence, India built upon its existing experience of using the Ganges River for irrigation, hydropower and flood control.

However, as the TWINS analysis showed, despite common goals to develop shared waters the specific norms and rules of water allocation were not easily finalized. The broader political context helps explain why. National and regional security issues were of major concern in the 1970s and 1980s, resulting in a tense diplomatic relationship between India and Nepal. Projects to develop shared waters became a useful geopolitical strategy for India as tensions in the Himalayan region grew in the 1960s. The Sino-Indo War in 1962 triggered by border disputes led to India actively seeking a cooperative relationship with Nepal to counter growing Chinese influence. Economic cooperation was a way to ensure Indo-Nepali alliance against China (Jha 1973: 39). This strategic move was reflected in increased Indian interest in bilateral hydraulic investment (Pokharel 1991: 82–83). Thus, water resources development became the cornerstone of 'good' neighbour relationships (Joint Communiqué in Dhungel and Pun 2009: 374). Large-scale projects, in particular those with hydropower components, were attractive, as they satisfied the strategic interests of the Indian government. The Power Exchange Agreement of 1971 bound Nepal to India, and prioritized projects like Karnali and Pancheshwar on the bilateral political agenda (see Joint Press Statement in Dhungel and Pun 2009).

The Nepali government considered India's military power to be a threat, especially when India intervened in several neighbouring countries, including Bangladesh, Sri Lanka and the Maldives, and when Sikkim was annexed (Pokharel 1991: 85; Upadhyay 1991: 33). To safeguard against any interference in domestic affairs, the Nepali monarch proposed the 'Zone of Peace' in 1975, an agreement which would restrict Indian military intervention in Nepal. However, in an attempt to exert its influence, the Indian government rejected this agreement and used other political opportunities to prevent Nepal from developing close ties with China. For example, in 1988 the renewal of the trade and transit treaties, which allowed landlocked Nepal to trade with other states via India, was delayed (Garver 1991; Murthy 1999: 1537).

This geopolitical context helps explain how issues over inequality in water allocation did not merely reflect the nature of the hydropolitical relationship between India and Nepal, but also the way in which national hydraulic missions were entwined with wider geopolitical concerns. The joint projects were, therefore, a medium through which the two states sought to control water resources and to impose a certain geopolitical order. India's hegemony is not only evident in the transboundary water domain but also in broader territorial and security issues in the region, thus attesting to Warner's (2008) claims about the multi-scalar aspect of hegemony and control.

Power asymmetry and defining 'mutual benefits'

The political economy and geopolitical dimensions of institutionalizing Ganges tributary development shed light on the attempt at and success of Indian hydro–hegemonic control of water resources. A further examination of power relations may be conducted through the discourse of 'mutual benefits'. The two governments had different interpretations of how benefits should be shared from bilateral projects. These interpretations were central to negotiations over large-scale projects.

For the Nepali hydrocracy, clear allocation of water resources and benefits became increasingly important for the assessment of hydraulic mission initiatives. The experience of the Kosi and Gandak agreements had a major impact on the negotiation strategies of elite decision-makers, based on the level of scrutiny water development projects received in domestic politics. The Kosi Project is viewed as one with a design that is far from optimal. There is criticism that the hydrocracy allowed the construction of the barrage too close to the international border so that India can divert as much water as possible, even though it would have been technically feasible to construct the barrage further upstream in Nepal (Nepali former governmental official A 2009, pers. comm.). With regard to the Gandak Project, it was claimed that India secured 97.2 per cent of the water from the new canals, leaving only 2.8 per cent for Nepal – thus skewing irrigation benefits in favour of the hydro-hegemon (Shrestha and Singh 1996: 90). In addition, the hydropower component took a long time to complete, leaving less opportunity for Nepal than India to exploit the hydropower capacity of the project (Pun 2009: 158).

These perceptions of skewed benefits contributed to shaping the discourse of equal water allocation for equal benefit sharing: benefit sharing 'on [a] 50:50 basis' (Minutes, April 1984 (Document 714) in Bhasin 2005: 1457). The Nepali discourse of mutual benefits promoted the norm of allocation based on equal proportion, taking into consideration existing water use. This is evident in negotiations about large-scale hydraulic projects between the late 1960s and the 1980s (see Sequence 3 of the TWINS matrix). During discussions over the Pancheshwar Hydro Electric Project, the Nepali delegation insisted that benefits, such as hydropower, irrigation and flood control, should be divided equally because the project was to be implemented on a shared, common river. It was argued that if one state consumes more than 50 per cent of these benefits, then it should compensate the other state (Minutes, December 1987 (Document 718) in Bhasin 2005).

However, this mode of allocation was fundamentally at odds with the Indian perspective. The discourse on mutual benefits presented by the Indian hydrocracy was based around the sharing of new benefits, arising from new hydraulic development opportunities. The Indian delegation negotiating the Pancheshwar Project made a directive speech act that politicized hydraulic development:

> [H]e [Naresh Chandra, Indian Secretary of Water Resources] stated that the sharing of power was not a problem because it was a commercial matter, but application of the same principle for water would not be appropriate or feasible. He emphatically expressed that payment of royalty for use of water in excess of 50% as proposed by HMG/Nepal could not be accepted and mentioned that acceptance of the equal sharing of water formula should not be made a precondition to implement the project.
>
> (Minutes, December 1987 (Document 718) in Bhasin 2005: 1483)

As a representative of the Indian hydrocracy, the Secretary of the Ministry of Water Resources claimed that mutual benefits accrue if projects are optimally designed and if they do not jeopardize existing water use (Record, June 1988 (Document 719) in Bhasin 2005: 1489). Furthermore, it was argued that projects involving non-consumptive water use – such as the Tanakpur project – should be excluded from the application of water-sharing norms (Minutes, December 1987 (Document 718) in Bhasin 2005: 1483). The Indian decision-makers interpreted Article 3 of the Mahakali Treaty to mean that the two states are entitled to equal water use of any *augmented* flow from the project (Iyer 1999: 1511). In short, while the Nepali hydrocracy argued for a clear and equal proportion of water resources and benefits, the Indian hydrocracy emphasized the importance of sharing benefits from the river that does not revisit existing water allocations of the basin.

The Pancheshwar Project negotiation continued to be an important agenda item when the Nepal–India Joint Committee on Water Resources (JCWR) was established in 2000. This committee is significant because it was the first secretary-level committee established to discuss transboundary water issues concerning all shared rivers. During its first meeting, the Indian Secretary of the Ministry of Water Resources emphasized the importance of cooperation and its eagerness to work with its Nepali counterparts (Minutes, November 2000 (Document 805) in Bhasin 2005). However, water allocation was not on the agenda for this high-level committee. In later JCWR meetings, Indian and Nepali representatives acknowledged the importance of the Pancheshwar Project, and other bilateral projects, and committed to establishing a project implementation agency. But, despite these discussions, no concrete decision was made to resolve the fundamental problem of the benefit-sharing issue (JCWR 2008, 2009). This demonstrates how the existence of cooperation, in the form of institutions such as a joint committee, does not necessarily clarify water allocation principles.

Moreover, it may be said that the JCWR put sanctioned discourse in place where only accepted issues can be discussed, and discourse that counteracts or challenges is not tolerated by the hydro-hegemon (Allan 2001, 2003; Zeitoun and Warner 2006). The Nepali view of water allocation and benefit sharing according to equal proportion receives no space to be aired in sanctioned discourse because water allocation in itself is not a legitimate discussion point within this organization. Thus, the indecision over the Pancheshwar Project reflects the tension over competing norms of water allocation, underpinned by India's hydro-hegemonic control.

Another hegemonic tactic employed was active stalling (Zeitoun and Warner 2006). The Indian hydrocracy exercised its comparatively stronger discursive power by persistently refusing to commit to the way in which benefits would be identified and shared. It was argued that project components such as 'the main dam and the re-regulating structure must be integrated for optimization of total benefits from the Project complex' (Summary record, November 1997 (Document 786) in Bhasin 2005: 1705). Discussing hydropower benefits was also postponed until the installed capacity of the hydropower component was determined (ibid.). In this way, water allocation would not be easily contested.

To counter Indian influence, the Nepali hydrocracy suggested outstanding issues be dealt with in technical expert group meetings, rather than through diplomatic channels (Letter, August 1997 (Document 783) in Bhasin 2005). This strategy to conduct the project details in the technical fora may have in fact allowed India to remain un-commissive and ultimately counter-productive. After 1997, much of the discussion among the Joint Group of Experts was highly technical in nature.[5] By engaging mostly in cooperative interaction in technical meetings, and focusing on collecting and assessing data, the key political decisions about the interpretation of the treaty were left untouched.

In addition, this discursive power of the hydro-hegemon was supported by the material capacity of the Indian hydrocracy. India's comparatively advanced economy, coupled with its experience in large-scale hydraulic projects, enabled it to plan and execute bilateral projects with Nepal. The Kosi and Gandak projects were essentially Indian projects, with most construction financed by India. In addition, the Indian hydrocracy was strongly supported by its government in its attempts to increase water allocation. The opportunitization of the Tanakpur project in 1992 (as exemplified in Sequence 4 in the TWINS matrix) would not have been possible had India lacked the financial resources to unilaterally construct the infrastructure. During the Pancheshwar Project negotiations, the Indian delegation was keen to avoid having to review existing water usage – even going so far as to suggest that it could finance the entire cost of irrigation water storage (Minutes, December 1987 (Document 718) in Bhasin 2005). Its aim was more to ensure equitable use of 'new' water consumption.

By contrast, the Nepali hydrocracy had less exploitation capacity (Zeitoun and Warner 2006). Not only were financial resources limited but there was also a dearth of infrastructure that could be used to control river flow and its benefits. Critics of Nepal's hydropower development policy claimed that, in order to make hydropower trade feasible, the power grid needed to be expanded and the power tariff mechanism revised (Bhadra 2004; Dhungel 2004). Moreover, the Nepali hydrocracy had little experience in the engineering of large-scale projects, leading to a reactive action rather than planning decisive strategies to secure benefits for Nepal (Nepali former governmental official A 2009, pers. comm.; Nepali former governmental official B 2009, pers. comm.; Nepali former governmental official C 2009, pers. comm.). Another intangible aspect of exploitation capacity is institutional memory. It has been suggested that opportunities to exert hydraulic control were lost because negotiators were unfamiliar with a project's background and the details of previous bilateral discussions (Nepali former governmental official A 2009, pers. comm.; Nepali former governmental official B 2009, pers. comm.; Nepali former governmental official E 2009, pers. comm.). In addition, domestic political instability was a stumbling-block, preventing effective planning and strategizing for water resources development (Nepali former governmental official B 2009, pers. comm.; Nepali former governmental official D 2009, pers. comm.).

In summary, Indian hydro-hegemony was achieved not through coercive means but rather with sanctioned discourse, active stalling and a strong resource base for hydraulic mission. This analysis provides more detail of the power asymmetry between India and its neighbouring basin states, as pointed out in the existing literature (e.g. Biswas 2011; Akanda 2012). The discursive strategies and material capacity are noteworthy in understanding the failure of the Pancheshwar Project. Consequently, bilateral institutions for cooperative water resources development cannot deal effectively with the issue of equity. Hydropolitical analysis, combined with explanations from the political economy of water resources development and geopolitics, shows that issues of inequality over water allocation and benefits are compounded by factors from the international and national scales of water resources management. This case study serves to show that explaining transboundary water interactions needs to go beyond the international scale of state-to-state negotiations.

Path dependency and the governance of the Ganges–Brahmaputra–Meghna River basin

What are the implications of the bilateral transboundary water interactions for the larger GBM River basin? This large river basin is a complex political landscape composed of six basin states: Nepal, India, China, Bangladesh, Bhutan and Myanmar. Basin-wide governance, which extends beyond specific projects, is argued in the literature to be a way of dealing with costs and benefits across the whole basin, especially as climate change impacts upon water availability at varying degrees in upper, middle and downstream areas (Crow and Singh 2009). High demand for energy is spurring many of the states in the GBM River basin to develop hydropower projects. In the Himalayan region of India alone, 292 dams have been proposed, and other Himalayan states are planning at least another 129 projects. This has raised concern about the absence of coordination between projects: transboundary planning is seriously lacking (Grumbine and Pandit 2013).

The track record of the transboundary water interactions between Nepal and India makes multilateral water governance challenging. These interactions focused on the use of key tributaries, and the institutional development of water use was conducted in a project-based manner. The implication could be that governance could be spatially patchy and not cohesive across projects. Moreover, the agreements failed to consider the environmental health of the rivers or resilience to potential environmental change, such as flow variability. The landmark Mahakali Treaty was not about sustaining water flow but rather about water use (Chakraborty and Serageldin 2004: 204). Other bilateral relations have not fared any better. Wirsing and Jasparro (2007) argued that transboundary water interactions between India and Bangladesh have seen some notable developments, such as the 1996 Ganges Treaty that determined water allocation at the Farakka Barrage. However, further discussions on the sustainability of the overall quantity of water flowing into the Farakka Barrage (especially during the dry season) has yet to be resolved. Moreover, with regard to the Teesta River, which contributes to the flow into the Brahmaputra River basin, the Bangladeshi hydrocracy claim that bilateral mechanisms have resulted only in multiple working groups that have little political influence or decision-making power (Wirsing and Jasparro 2007: 236–237). This situation has allowed India to actively stall decisions on water sharing.

To explore this question about the link between bilateral transboundary water interaction and basin-wide governance, I draw on the concept of path dependency. This concept helps explain why institutions have focused on a narrow set of criteria for water allocation. Path dependency may be understood as an 'organizing concept' that explains the temporal process of policy development (Kay 2005: 554). This concept places the process of decision-making into its historical context, from which constraints upon future decisions and actions are derived. Heinmiller (2009) argued that path dependency is developed through vested interests, sunk costs, formal/informal contracts and network effects. These factors restrict the project uptake of institutions focusing on conservation and sustainable water use. The consequence is that institutions which focus on water allocation for economic development are securely put in place and become hard to change. Water allocation institutions invest in hydraulic infrastructure to facilitate allocation. Because these are capital-intensive investments there is high incentive for the continual usage of the infrastructure, in order to recover sunk costs. Formal/informal contracts can establish rules – such as property rights – thus stabilizing water allocation. Network effect refers to the integration of rules at multiple governance levels; once water allocation rules are integrated across scales, it becomes difficult to change them (ibid.: 135–136). Furthermore, this process increases transaction costs, prohibiting a shift to any new type of arrangement (Kay 2005).

With regard to the institutionalization of water resources development between Nepal and India, sunk costs accrued through the national hydraulic missions, which invested heavily in barrages, canals and hydropower components. The bilateral agreements, as formal contracts between the two governments, had the effect of establishing rules of water use. Even though it would have been possible to make revisions later, the initial negotiations had already taken a long time, leading to high transaction costs. The network effects of the institutions may also be observed in the number of committees established at various levels: the Kosi Co-ordination Committee; the Gandak Co-ordination Committee; and JCWR. These committees established a formal governance framework which allowed for the creation of further organizations such as the Standing Committee on Inundation Problems (later, the Joint Committee on Inundation and Flood Management) and the Power Exchange Committee. The path dependency of bilateral agreements governing projects on the Ganges tributaries constructed a decision-making process that prioritized water for economic development. Moreover, through the development of institutions, the logic of the construction of large-scale projects providing economic benefits was unquestioned. Path dependency suited the interests of the hydrocracy in India, and in Nepal.

Path dependency thus embodies an inherent bias on how water resources should be managed. In the Ganges River basin, bilateralism over water issues has been the norm because the Indian government prefers to have bilateral foreign policy strategies rather than multilateral ones (Crow and Singh 2000: 1910–1911; Brichieri-Colombi and Bradnock 2003: 50). Crow and Singh (2009: 320–321) explained that the Indian government considered bilateralism as a necessary and better strategy for economic development in the South Asian region. Transposed to the context of water resources, it was understood that bilateral discussion provides prompt decision-making. In contrast, the non-hydro-hegemonic countries Nepal and Bangladesh have viewed bilateral approaches as prohibiting collective bargaining that could counter India's control over water resources (Crow and Singh 2000: 1910–1911). As the previous section showed, India employed hydro-hegemonic tactics of active stalling through bilateral agreements. Contending with power asymmetry is necessary for basin-wide governance.

The practice of bilateralism poses a challenge to the multilateral governance of the larger GBM River basin. Some efforts have been made to achieve regional cooperation. However, they have so far been limited. The South Asian Association for Regional Cooperation (SAARC) – formed by the Maldives, Sri Lanka and the GBM River basin states (excluding China) – has avoided specific discussion on transboundary waters. As an initiative distinct from SAARC, the South Asian Growth Quadrangle (SAGQ) formed in 1997 was supported by Nepal, India, Bangladesh and Bhutan, with the aim of facilitating transportation and trade, which has implications for river navigation and hydropower development. However, even at its inception, analysts such as Hossain (1998) identified obstacles, such as the lack of political will to make it work. Despite the Asian Development Bank (ADB) promoting SAGQ, very little work has been carried out. In the 1990s, the Ford Foundation supported leading policy institutions in Nepal, India and Bangladesh to provide some leadership. The Institute of Integrated Development Studies of Nepal, the Centre for Policy Research of India and the Bangladesh Unnayan Parishad convened several meetings to discuss potential options for water cooperation and produced joint publications on this topic (e.g. Verghese *et al.* 1993; Adhikary *et al.* 2000). However, how the discussion within these initiatives has led to change in water allocation negotiation is far from tangible.

Path dependency also perpetuates the geographical imagination of the shared river basin as the stage for engineering projects, further lending support and legitimacy to the hydraulic mission. The Indian government has pursued the idea of a country-wide project to transfer water across long distances in order to meet water demand. Commonly referred to as the River Linking Plan, the National Perspective Plan for Water Resources was originally proposed in 1980. The idea was to divert water from water-rich regions in the Himalayas to, for example, the Southern Indian peninsula for better irrigation opportunities. The River Linking Plan identified the Kosi, Gandak, Mahakali and Karnali Rivers, as well as water from the Brahmaputra basin, geographies of water abundance suited for large-scale transfer. The Indian government disclosed that the irrigation potential could be increased by 35 million ha, and 34,000 MW could be generated in this way (Government of India 2002). The proposed hydropower development throughout the GBM River basin is also a reflection of the geographical imagination, of a river basin full of opportunities for development. This imagination gives rise to a competitive climate of river basin development that makes basin-wide governance challenging and even impossible. The Chinese government is now seeking to exploit the Brahmaputra basin more actively than before, planning dams for energy production and water transfer for irrigation (Chellaney 2013). As China begins to look towards its borders for 'untapped' water resources, the pressure is mounting on the Indian government to respond by accelerating the River Linking Plan (Pearce 2012).

New forms of cooperation in the Ganges–Brahmaputra–Meghna River basin?

Multilateral initiatives over water have faced political sensitivities. The Strategic Foresight Group, mandated 'to building peace and security using water as an instrument for cooperation between countries having difficult relationships', had very low expectations of high-level political commitment between China, Nepal, India and Bangladesh over shared waters (Strategic Foresight Group 2011: 1). The possibility of ministerial-level meetings was discounted because the governments would not buy into dialogue that contravened existing policy (ibid.). However, driven by increasing energy demands and climate change concerns, the South Asian Water Initiative (SAWI) is an example of efforts at better governance among Bangladesh, Bhutan, China, India, Nepal, Pakistan and Afghanistan. Led by the World Bank and with British, Australian and Norwegian donors, SAWI has demonstrated its convening power by including a large number of countries, including China. One of its major outputs from this initiative has been the Ganges Strategic Basin Assessment (World Bank 2010a). This assessment uses basin-wide models to examine hydrological and economic implications within different climate change and development scenarios. Addressing perceptions about the utility of upstream storage for managing downstream flow (especially during dry seasons) is a key aim, and the empirical data provide a base for discussions on hydropower capacity (World Bank 2010a, 2012). Knowledge production of this kind may be helpful because it provides tangible indicators of development options. However, as Brugnach and Ingram (2012) lucidly noted, it is not just the data and information on climate change and development that is necessary but also the multiple and diverse types of knowledge to communicate and deliberate the different views in a decision-making process.

This point about knowledge co-production is pertinent, especially because there are claims that new forms of regional cooperation are required to manage increasing hydropower development under conditions of climate change and its transboundary environmental impact (e.g. Ebinger 2011; Matthew 2013; Shah and Giordano 2013; Rasul 2014). Simply having new cooperative initiatives is not sufficient. This is especially so when decision-making is being influenced by private investors and power producers that engage the hydrocracy and other elite decision-makers in pursuing the hydraulic mission. Regional initiatives require ways of mediating the divergent interests of stakeholders, not to mention culture and rationalities. Otherwise, while multilateral initiatives – especially those with external funding – may be tolerated to an extent, water resources will continue to be used as instruments for increased agricultural productivity and hydropower development.

Notes

1 India became independent in 1947.
2 The latter also stretched into Nepali territory, and the infrastructure within its territory would also be the ownership of the Nepali government. This section of the infrastructure was called the Eastern Nepal Canal.
3 The Sarada Barrage fed water to the Sarada Canal Project. It is said that the Tanakpur Barrage would continue to facilitate an Indian irrigation project with capacity for 1.61 million ha (see Pun 2009b).
4 See the scale of protest in reports such as Xinhua General News Service (1992, 1993a, 1993b).
5 See e.g. Documents 792, 794, 795, 799 in Bhasin (2005).

5 Engineering the Orange–Senqu River

Introduction

The Orange–Senqu River basin has several unique features, making it an important basin in Southern Africa. With a basin area of approximately 1,000,000 km² across Lesotho, South Africa, Botswana and Namibia, and a natural runoff of approximately 11,300 million m³ per annum, it is the largest river basin in the region (ORASECOM 2014) (see Figure 5.1). Hydropower generation is a major function of this basin, to the extent that it is reported that it alone supplies approximately half of the African continent's electricity (GEF 2012: 15). Furthermore, it has been the setting for one of the world's large-scale international transboundary water transfers. Within the upper basin, the Lesotho Highlands Water Project (LHWP) transfers water from Lesotho to South Africa for industrial use and to provide potable water. At the same time, this project poses issues of water allocation downstream where the river flows through more arid areas. Extensive measures to use and abstract water resources have been put in place over time. Analysis of transboundary water interactions over access and allocation of 'non-food water' (Allan 2013), or for industrial and domestic purposes not related to agricultural activity, is examined with the LHWP case study.

This chapter focuses initially on the relationship between Lesotho and South Africa during the development of the LHWP. Currently, 30 m³/sec is transferred between the two states and while the annual transfer of less than 1,000 million m³ (1 km³) may be small in quantity, it is of high economic value. This is because it provides water to key urban centres such as Johannesburg and Pretoria, which are located away from large blue water sources. The decision-making of access and allocation has engaged a group of elite decision-makers from the LHWP project authorities and governmental officials from key ministries, heads of state and politicians. Currently, a joint commission, the Lesotho Highlands Water Commission (LHWC), governs the project.[1] In Lesotho, the hydrocracy revolves around the Ministry of Natural Resources and the project authority, the Lesotho Highlands Development Authority (LHDA). In South Africa, the hydrocracy is centred on actors from the government agency involved in water issues: the Department of Water Affairs (DWA), later renamed the Department of Water Affairs and Forestry (DWAF) in 1990 and reorganized to the DWA in 2009. The project authority, the Trans Caledon Tunnel Authority (TCTA), is also a key organization that has driven river basin planning, with a large stake in the development of the LHWP.

Analysis of the interaction of these actors unpacks the ways in which conflict and cooperation occur between Lesotho and South Africa. In addition, it becomes evident how this bilateral transboundary water interaction makes up basin-wide water governance that attempts to address issues of water access and allocation at different spatial scales. Multi-scalar water governance brings about interesting dynamics among the four basin states. The four basin states are differently endowed and engage in water management institutions at the regional and basin levels, in addition to bilateral institutions. Lesotho is an upstream state, with the headwaters of the Senqu River – one of the two main tributaries of the Orange–Senqu River – within its territory. South Africa has been actively planning and developing projects that use the other main tributary, the Vaal River. This state is both downstream vis-à-vis Lesotho and upstream vis-à-vis Namibia. Botswana does not contribute to the perennial flow (Heyns 1995: 479), but it shares other international transboundary river basins with

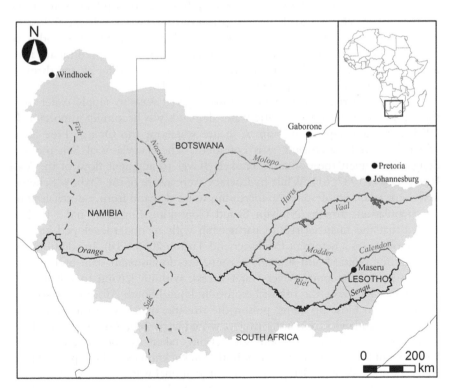

Figure 5.1 Map of the Orange–Senqu River basin (source: author).

South Africa and Namibia such as the Okavango, Limpopo and Zambezi. I argue that the notion of space and territory becomes central to the consideration of benefits from water sharing and solutions to water scarcity. While there are basin-wide and regional initiatives to enhance multilateral cooperation over water, analysis shows that these initiatives are not effective at countering national interests and hydro-hegemonic control. Examination of transboundary water interactions shows that regional benefits are an elusive concept.

Seeking potential in the Senqu River

The water flow between the two tributaries in the upper sub-basin of the Orange–Senqu River basin differs only slightly. The Senqu River releases 4,730 million m³ per year and the Vaal River 4,270 million m³ per year (Earle *et al.* 2005: 1). However, the small basin area in the mountainous region of Lesotho contributes to 41.4 per cent of the mean annual runoff (ORASE-COM 2014). These conditions of the Senqu River led to the idea of a water transfer project, which takes advantage of relative water abundance. British powers, which ruled the protectorate state of Basutoland until its independence as Lesotho, took the first step in assessing the potential. The High Commission and Director of Public Works of Basutoland conducted an initial study in 1950. This led to the investigation of the Oxbow Scheme following independence in 1966 by the newly formed government of Lesotho to transfer water from the Senqu to the Vaal River basin. The preliminary feasibility studies developed in 1967 drew up plans to transfer water through tunnels and generate hydropower (DWA 1987: 61).

The South African hydrocracy also investigated ways to supply water from the sub-basin, as meeting growing water demands was becoming important. In 1955, using the upper sub-basin to secure water for the Orange Free State, which bordered Basutoland, was considerd. The Tugela–Vaal Scheme was designed to divert the domestic Thukela River to augment flow of the Vaal River (DWA 1987: 61). While hydrocracy were aware of the Oxbow Scheme, it was not taken up, despite the increasing water demand from industrialization and population growth (Ninham Shand Consulting Engineers n.d.: 4). It is argued that the financial risk of partnership with an underdeveloped country like Lesotho was a deterrent (Meissner and Turton 2003: 118). In addition, it is said that for political and strategic reasons, the downstream state preferred to have independent control over water access, without having to rely on the other state. This way, the risk of Lesotho 'hold[ing] South Africa to ransom by threatening to blow up dams, poison the streams, or sabotage the supplies in other ways' with a water transfer project would be avoided (Simons 1968: 138).

This series of individual domestic actions, taken by the governments of Lesotho and South Africa, show that both states acknowledged the possibility of water transfer. Until the late 1960s, the international water transfer scheme was not discussed formally between the two states, making it a non-politicized issue on the bilateral agenda (TWINS Sequence 1 in Figure 5.2). It is important to

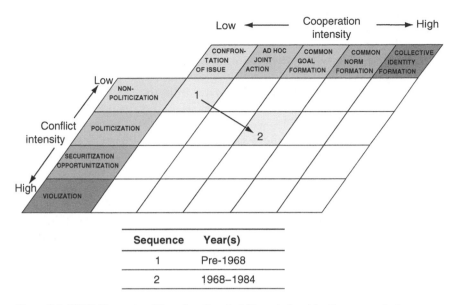

Figure 5.2 TWINS matrix of Lesotho–South Africa relationship: Sequences 1–2.

Note
Numbers in the matrix represent sequence of significant speech acts. The position of sequence numbers within a cell is for presentation purposes only and it does not represent degrees within a certain conflict or cooperation intensity level.

note that while Lesotho has better geographic access to headwaters, using 'abundance' was limited to one option of water transfer which required the interest of South Africa to materialize. In contrast, the South African hydrocracy had the means to consider multiple plans and to reject the Oxbow Scheme. This signifies South Africa's exploitation potential (Zeitoun and Warner 2006), putting it in a hydro-hegemonic situation compared to Lesotho.

Bilateral negotiations and the changing geopolitical context

The water transfer plans entered the bilateral agenda when droughts in the mid-1960s triggered the South African hydrocracy to look for alternative water resources. The Oxbow Scheme became a realistic option as the DWA considered the use of the Upper Orange–Senqu sub-basin waters. This way, it would be possible to augment flow of the Vaal River and allow storage in the Vaal Dam, crucial for irrigation (DWA 1987: 61). As a result, an agreement in principle on water transfer was reached between the two governments in 1968 (Meissner and Turton 2003: 119). This commissive speech act signifies the formation of common goals to develop hydraulic infrastructure for water transfer (TWINS Sequence 2 in Figure 5.2).

However, the procedure of agreeing details of the water transfer was disputed, politicizing the project. First, the two countries were in dispute over the mechanism for royalties for water transferred, to the extent that negotiations were interrupted in 1972 (Mohamed 2003: 225). In the end, a comparative approach to determine royalties was adopted. Cost savings which South Africa would make by receiving water from the LHWP was calculated based on an alternative domestic water transfer, the Orange Vaal Transfer Scheme (OVTS). The OVTS would require infrastructure to pump water from the Orange River to the Vaal and thus have higher costs than the LHWP, which uses gravity flow for water transfer (DWA 1986; Wallis 1992). This comparison provided grounds upon which fixed royalties could be determined, in addition to variable royalties for the actual water transferred.

Second, identifying the project parameters was time consuming. A joint commission with Basotho and South African representatives formed the Joint Technical Committee to undertake a joint preliminary feasibility study in 1978 (LHDA–TCTA 2001: 1), but it was only in 1986 that the final feasibility study was completed. Despite it being a joint commission, the Basotho and South African delegation operated separately and met only to review each other's work owing to funding restrictions (DWA 1987: 63). A full joint feasibility study was undertaken between 1983 and 1986 detailing the engineering requirements, legal and institutional arrangements for water transfer, and the timeline for implementation. The scale of the project required new and extensive information, making it 'the most comprehensive water resource planning investigation in the history of Southern Africa' (DWA 1989: 3). For the South African hydrocracy, having multiple options for water supply was important. Consequently, they engaged in lengthy negotiations over the water transfer project and used resources to investigate various options. The project was thus framed as one of the technically viable solutions to manage water scarcity.

Moreover, negotiations were interrupted as diplomatic relations deteriorated between the two states. This process of determining project details was held in the wider context of increasing regional instability, as apartheid and Cold War issues divided South Africa and its neighbouring states. The LHWP seemed initially to fit the South African government's stance on 'political independence amid economic interdependence' but hostility increased during the 1970s (Geldenhuys 1982: 132). The South African government labelled Lesotho 'an extremist state' that supported anti-apartheid movements through the African National Congress (ANC), a political party that was banned in South Africa at the time (Meissner and Turton 2003: 120, citing Barber and Barratt 1990). In turn, the Basotho government accused South Africa of supporting the anti-governmental organization, the Lesotho Liberation Army, which had attempted to destabilize the government by carrying out attacks on infrastructure, such as small hydropower generators in Lesotho (Mochizuki 1998: 135). Moreover, the Basotho government, in seeking ties with communist countries such as the Soviet Union, China, North Korea and

Cuba, was seen to aggravate the relationship, especially as communist influence was regarded as a serious threat to South Africa (Jaster 1988). Progress on the LHWP had stalled and protracted by the end of the decade, though commissive speech by heads of state in 1980 reassured commitment to the project, and again in 1983 (BBC 1980, 1983; Turton *et al.* 2004: 244).

Securitizing water transfer

The geopolitical climate provides the backdrop to the process of securitization of water transfer that firmly embedded the LHWP in South African security policy. In 1984, the South African Minister of Foreign Affairs, Pik Botha, threatened to abandon the project if Lesotho did not sign a security pact to contain further hostility (Rand Daily Mail, 1984). This directive speech act was instrumental in socially constructing the LHWP as an object of national security. In other words, the project was 'dependent on satisfactory arrangements to secure [it] from sabotage' (Laurence 1984: 4), without which there could be no progress with bilateral negotiations. This claim was followed up with South African technicians temporarily withdrawn from the project (Meissner and Turton 2003: 120). As a securitizing practice, the South African government blocked the border with Lesotho in January 1986, claiming this was necessary to protect its territory from terrorists. Lesotho, being landlocked within South Africa, quickly faced a major economic crisis as it was cut off from basic and important resources, including food (Braun 1986).

The discourse presented by the South African government portrays a situation where, at this time of crisis, the joint project can only succeed by tabling a security agreement: no other measure would suffice. The border blockade is treated as a necessary and timely intervention to contain further crises. This discourse emphasized urgency and emergency measures that would not normally be taken, an element crucial to securitization (Buzan *et al.* 1998; see also Chapter 3, this volume). Ultimately, this speech act aimed to win Lesotho's compliance by framing the LHWP as one relating to state survival within an unstable region (TWINS Sequence 3 in Figure 5.3).

The Basotho government had maintained that the water transfer project was completely separate from security issues. Rather, it presented an opportunity for economic cooperation (Freimond and Pitso 1984). However, in this extreme situation of border blockage, a military coup d'état occurred, ousting the democratic government and replacing it with a new regime led by a military organization. This change of government in Lesotho triggered a series of events that reduced conflict intensity of transboundary water interactions from securitized to politicized. The new head of state, Justin Metsing Lekhanya, made assertive speech acts by declaring his intent on proceeding with the project:

This [Lesotho Highlands Water Project] has been a national dream for all Basotho and the Republic of South Africa. The new government will respect all bilateral and multilateral treaties entered into by the previous government. And the Highlands water project is not excluded. We are aware of the enormous economic potential the project has for this nation with regard to hydroelectric generation, irrigation, and the much-needed income to finance government operations, and we do hope we are continuing to negotiate with South Africa on the project, and whatever differences exist, we shall iron them out on the negotiating table, and in line with our general policy in trying to get this project into being. We are very keen to see this materialising and coming into being.

(BBC 1986a)

For the Basotho hydrocracy and elite decision-makers, the project was economically attractive. In an interview shortly after the establishment of the new government, the head of state stressed that the project has 'enormous economic potential for this nation with regard to hydroelectric generation, irrigation, and the much-needed income to finance government operations' (BBC 1986a). The LHWP was framed as a long-sought-after solution to using abundant water for economic benefits. The new government won domestic support as well as the confidence of the South African government (Cowell 1986; Mills 1992; Hayashi 1999). These changes speeded up the

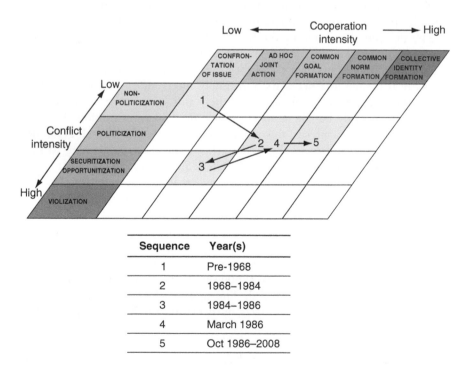

Sequence	Year(s)
1	Pre-1968
2	1968–1984
3	1984–1986
4	March 1986
5	Oct 1986–2008

Figure 5.3 TWINS matrix of Lesotho–South Africa relationship: Sequences 3–5.

negotiations, with the South African Foreign Minister emphasizing in March 1986 that a formal agreement was imminent (BBC 1986b).

The full feasibility study found that there were no major engineering obstacles to the storage of water in the Lesotho highlands or to the transfer of water, giving it the green light (Lahmeyer MacDonald Consortium and Oliver Shand Consortium 1986). In October 1986 negotiations resulted in the signing of the *Treaty on the Lesotho Highlands Water Project between the Government of the Republic of South Africa and the Government of the Kingdom of Lesotho* (hereinafter Treaty on the LHWP). The commissive speech act of signing the treaty indicated a commitment to construct and implement the water transfer project. Furthermore, it formed the norms of water transfer, as explained in detail below (TWINS Sequences 4–5 in Figure 5.3).

Some analysts speculate that the South African government played a role in the coup d'état in order to achieve agreement on the water transfer project (Homer-Dixon 1999; Matlosa 2000; Furlong 2006). These studies attempt to show the causal linkage between low and high politics. Indeed, this securitizing practice seems to have been the most overt and severe action employed by South Africa to apply economic pressure on neighbouring states within this tense geopolitical setting (Jaster 1988: 135). However, concrete evidence showing a causal relationship between South African intervention and the outcome in Lesotho is hard to come by in the public domain. In other words, whether the environmental issue of water supply, a low politics issue, was subject to high politics within the national security and geopolitical agenda is difficult to ascertain, much less explain whether it drove development in thigh politics (Mirumachi 2008: 44–45). Moreover, distinguishing between high and low politics may not be very helpful either. Securitization has the effect of blurring the lines between low and high politics, disregarding the problem characteristics because of the way in which social construction can frame an issue as a crisis. The restricted group of actors participating in decision-making also facilitates this process of '*silence, secrecy* and *suppression*' (Roe 2006: 426; emphasis in original). In the case of the water transfer project, the hydrocracy in both Lesotho and South Africa had a strong interest in using the water, and the politicians of the two states were attracted to the potential economic advantage. Rather than guessing at the possible politics and bargaining surrounding the coup d'état, analysing the securitizing discourse and practice gives a better indication of the entrenched interests of the elite decision-makers and their strategy for securing benefits from joint development.

Framing the success of the LHWP

Guided by the bilateral treaty, construction began to engineer the river flow. Key infrastructure included the 185 m-high Katse Dam with its storage capacity of 1,950 million m³, the Mohale Dam with its maximum capacity of 947 million m³, and the Matsoku Weir, all constructed in Lesotho. The first phase of the project was implemented in two stages with Phase 1A completed in 1998 and Phase 1B in 2004. The infrastructure completed during the first phase of the project allowed water transfer at the rate of $30\,m^3/sec$. In addition, the first large-scale hydropower plant in Lesotho, the 72 MW Muela Hydropower Plant, was also constructed during Phase 1B. These physical changes to the river basin represented visible signs of progress in the eyes of elite decision-makers. Their assertive speech acts underscore a discourse of successful bilateral cooperation. For example, political figures such as King Letsie III of Lesotho claimed:

> In this world class example of cross-border cooperation, there are only winners.... Lesotho benefits economically from the royalties.... South Africa will ameliorate its water supply problem.
>
> (Reber 1998: n.p.)

The South African President, Thabo Mbeki, highlighted the economic benefits for the two states:

> The Lesotho Highlands Water Project is a bi-national project to harness a natural resource, Lesotho's 'white gold', for the benefit of both our countries. For South Africa, the project brings improved security of water supply for both economic and domestic use, and will undoubtedly help to meet the increasing water demand for many years to come. Equally, Lesotho enjoys the benefit of new infrastructure including roads, expanded communication and electricity systems, health facilities, job opportunities, improved water supply and sanitation to numerous communities and many additional secondary benefits associated with a huge capital investment with its revenue streams.
>
> (Mbeki 2004: n.p.)

These assertive speech acts highlight (1) the intention to achieve economic benefit from the cooperative interaction, and (2) the belief, by the Basotho and South African heads of state, that the project was a suitable vehicle for realizing such benefit.

Public ideas (Ringius 2001; see also Chapter 3, this volume) related to the hydraulic mission facilitated the construction of this economic development discourse. In other words, large-scale engineering solutions addressed the problem of water scarcity. Speech acts reveal that water transfer was considered to be the 'rational' or 'efficient' way of managing water resources by both the Basotho and South African hydrocracies and politicians. The discursive framing of the LHWP as an enterprise for economic benefit is also reflected in the institutional development of water transfer, best examined through the treaty's common norms. Specifically, a clear mechanism of royalty payments was set up to maximize the economic benefit of developing the Senqu waters (Boadu 1998). The South African government pays fixed and variable royalties to Lesotho. As mentioned earlier, the fixed royalties are calculated on a comparison with a domestic water transfer project in South Africa. The savings made by implementing the LHWP amount to 56 per cent of the cost of the domestic OVTS (Wallis 1995: 7). Variable royalties are based on each cubic metre of water transferred. Annex II of the treaty makes provisions for year-on-year minimum water delivery quantities. The costs of the project are also laid out in the treaty. All construction costs for water transfer, as well as the costs associated with operation and maintenance of the transfer, are the responsibility of South Africa. In contrast, it was agreed that the government of Lesotho would fund any ancillary development. The treaty emphasizes that the project would be guided by 'market efficiency principles, leaving little or no room for political prices that were inconsistent with market efficiency' (Boadu 1998: 401). Rules for identification of costs are included in detail in the treaty, signifying the priority given to the economic value of the project.

Reports indicate that following completion of the first phase of the project, the LHWP contributed 14 per cent of GDP and 28 per cent of government revenues for Lesotho, with over 15,000 jobs directly generated (TCTA–LHDA 2003). Using figures of revenue between 1996 and 2010, one million cubic metres of highlands water was worth an average of R0.38 (LHWP 2010). In addition, annual fixed royalties during Phase 1 were estimated to be approximately US$55 million (LHDA n.d.a: 20). Ancillary development projects have also changed the infrastructural landscape of Lesotho. As part of the project, roads into the highlands, clinics, schools and communication infrastructure were developed. However, it is important to point out that framing the project as a major economic opportunity was challenged by environmental activists and those concerned with its impact upon livelihoods. Local and international NGOs pointed out the trade-offs this 'development' had cost. Environmental impacts have been reported: for example, downstream deterioration of water quality and loss of fish diversity (Hoover 2001; Lepono et al. 2003); loss of flora and fauna, including traditional medicinal herbs used by the local communities in Lesotho (Hoover 2001); and seismic activity (Nthako and Griffiths 1997; TCTA–LHDA 2003). Negative socio-economic impacts have also been raised, particularly in relation to livelihoods, and inadequate compensation plans and development initiatives for the resettled Basotho communities (Horta 1995; Archer 1996; Hoover 2001, Scudder 2005).[2]

Local NGOs, such as the Transformation Resource Centre and the Christian Council of Lesotho, raised issues on behalf of the local communities from an early stage. These organizations often joined forces with regional and international NGOs, such as the International Rivers Network, Environmental Defense and Christian Aid. In addition, South African civil society was active in raising concerns. The Development Research Institute acted as a representative for over 100 civic associations, enabling them to protest against further development plans – their argument based on the mounting evidence of detrimental environmental impact and inflated water prices (WCD Secretariat 2000: 84–85). The South African Minister of Water Affairs and Forestry, Kadar Asmal, publicly acknowledged that there had been an adverse impact on the local communities, and that measures designed to mitigate such an impact had not been executed in a timely manner (Salgado 1998). Lessons learned from Phase 1A were taken into account to some degree. For example, the Environmental Action Plan of Phase 1B was conducted *prior* to construction, and an Instream Flow Requirement report was carried out to ensure downstream flow (TCTA–LHDA 2003: 14–15). In addition, during Phase 1B, a more comprehensive compensation package including greater choice of compensation methods and larger financial support was offered to local communities.

None the less, the LHWP was framed as an important economic development measure. Claims by the project authorities about the potential loss to the economy, in the region of R1.5 billion to R8 billion per annum, helped strengthen the need to address risks of water scarcity with the LHWP (TCTA–LHDA 2003: 44). The project was also couched in terms of model cooperation between the two countries. For example, the South African President, Nelson Mandela, asserted in 1998:

> The stage is set for sustained development and growing prosperity. It is the promise of a better future for our children.... The resounding success of the Lesotho Highlands Water Project testifies to the powerful spirit of cooperation that is growing as Africa lifts itself through its own efforts.
>
> (Salgado 1998: n.p.)

The head of state in Lesotho also emphasized success: 'The project has indeed served well as an international example of co-operation between the Kingdom of Lesotho and the Republic of South Africa' (King Letsie III 2004: n.p.). These speech acts demonstrate the disjuncture between the interests of the hydrocracy, backed by the powerful political elite, and the local communities. Bilateral cooperation during the initial stage of the project was shaped largely top-down. Decisions and implementation through formal institutional mechanisms, such as RBOs, protocols and agreements, did not serve well in reflecting environmental and socio-economic concerns mounting at the local community level.

This is not to say that the project was depoliticized. Rather, the process of implementing the norms of the project was contentious. For example, there was a dispute about the use of hydrological data to calculate the net benefit of water transfer and royalties. In addition, the interpretation of the treaty clause regarding project cost was also disputed. Specifically, the disagreement was about whether South Africa was responsible for income tax payment as part of the project costs. A compromise was eventually reached that meant Lesotho could levy taxes for economic activities related to the project, and South Africa would be responsible for their payment, but only to a certain upper limit (Tromp 2006: 48–49). These issues arising from the details of the project are important, in that they built up a body of protocols to institutionalize project development. The particular issue of income tax payment was institutionalized in Protocol V in 1999. Another arrangement, Protocol VI, is particularly significant because it established the bilateral Lesotho Highlands Water Commission (LHWC). This new governance mechanism replaced the former Joint Permanent Technical Commission (JPTC). The LHWC was tasked to monitor the two national project authorities representing Lesotho and South Africa: the Lesotho Highlands Development Authority (LHDA) and the Trans Caledon Tunnel Agency (TCTA). The LHWC was particularly effective in providing some level grounds of negotiation between the two governments, especially as it was observed that the South African delegation was reluctant to accept development policies and attempted to 'reduce, and in some cases eliminate' their responsibilities (Scudder 2006: 53).

The relatively high level of cooperation coexisting with a low level of conflict, as exemplified in Figure 5.3, is a reflection of a strong discursive framing of the LHWP by decision-makers underpinned by institutional development where the LHWC became the vehicle through which norms were interpreted. Protocols and agreements thus played an important function in the institutional development of bilateral water transfer. How this institutional development influenced, and was influenced by, other institutions to govern water at different spatial scales will be discussed in the second half of this chapter.

Stalled project expansion

As the first phase came to a close, the economic value of the project became an increasingly sensitive issue for the basin states themselves. At the time of the Phase 1 agreement, it was envisaged that the project could be expanded in scope by an additional three phases, with the aim of transferring a total of 70m³/sec. According to Article 6 of the Treaty any subsequent phase must be subject to prior consent. Thus, new political negotiation was needed before any further phases could be implemented. In 1999, negotiations regarding the second phase began (SAPA News Agency 1999). Initial plans suggested that Phase 2 would be operational by 2010 (Wallis 1992: 38). However, it was not until 2004 that an initial agreement to conduct feasibility studies was reached, and it took another year to finalize (King Letsie III 2004; Mbeki 2004; DWAF 2005) (TWINS Sequence 5' in Figure 5.4).

The South African government was hesitant about proceeding with the second phase immediately. Instead, the hydrocracy was mobilized to investigate the Thukela Water Project (TWP), a domestic project to divert water from the Thukela River located in KwaZulu-Natal to the Vaal River. This alternative project was investigated against the background of increased awareness about LHWP Phase 1 impacts, and more scrutiny of environmental and socio-economic impact from civil society. In contrast, the Basotho

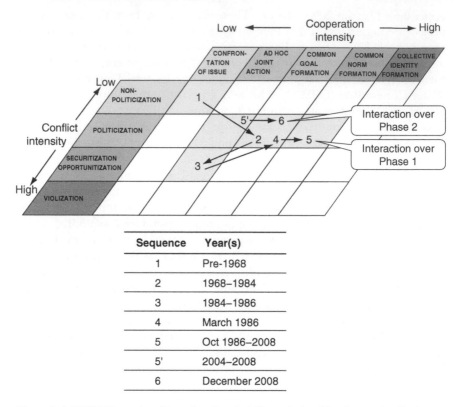

Figure 5.4 TWINS matrix of Lesotho–South Africa relationship: Sequences 5'–6.

government was keen to proceed with the next phase, claiming that it would further reduce poverty. The Minister of Natural Resources, Monyane Moleleki, made a directive speech act in favour of the second phase:

> From the government of Lesotho standpoint, the Lesotho Highlands Water Project is of strategic importance in ensuring poverty alleviation in Lesotho.... In my opinion, Phase II of the Lesotho Highlands Water Project would optimise the current Phase I infrastructure. Consequently it must be the cheapest option for the Republic of South Africa. The consultants and professionals who will represent the Lesotho Group on the feasibility studies must be very careful and ensure that the options proposed will maintain Lesotho's competitive advantage over *any* alternative looked at by the government of South Africa.
>
> (LHWC 2004; emphasis added)

The Basotho government urged South Africa to commit to the second phase as soon as possible, but was frustrated by the slow progress (Basotho governmental official 2006, pers. comm.). Even when feasibility studies for Phase 2 were completed DWAF was non-committal, and the Deputy Director-General emphasized that various options remained subject to investigation. This assertive speech act was used to 'keep options open' on the LHWP, in order to ensure a better bargaining position (Guvi 2008). As an indication of the sensitivity of the project, during an interview for this research a member of the hydrocracy from one of the basin states remarked that publicly commenting on the topic is risky to the extent that it could cost his or her job. While joint action to conduct feasibility studies was made, Phase 2 was politicized and the nature of the common goal to achieve further project development disputed.

In December 2008, the South African government announced that the second phase of the project would be implemented. The agreed Phase 2 involved the construction of the Polihali Dam, with a capacity of 2.2 billion m³ (ORASECOM 2013a). This dam was significantly smaller than the Mashai Dam (3.3 billion m³), proposed at the time of the Phase 1 agreement (Orange–Senqu River Awareness Kit 2010) (see Figure 5.5). This initial agreement was a commissive speech act of the two governments that signified the intention to engage in joint action, and established the shared goal of increased water transfer. This joint action and shared goal make up the

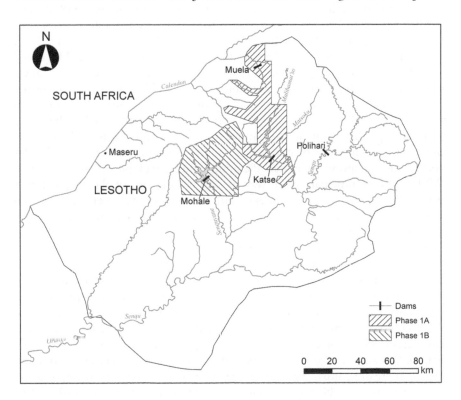

Figure 5.5 Map of project sites for phases 1 and 2 of the LHWP.

TWINS cooperation intensity of common goal formation. At this point in 2008 no formal agreements were signed, meaning that cooperation intensity was short of 'common norm formation' according to the TWINS matrix (TWINS Sequence 6 in Figure 5.4). Common norm formation would require further negotiations of the protocols specific to Phase 2 to fulfil the rights and obligations of the LHWP treaty. The temporal scope of the TWINS analysis of this case study extends until the point of initial agreement of Phase 2. This case study demonstrates how several conflict and cooperation trajectories may be analysed depending on various issues within bilateral trans-boundary water relationships, in this case on Phase 1 and Phase 2 of the LHWP.

The hydrocracy and geographies of hydraulic development

For hydrocracies of Lesotho and South Africa, the LHWP was an important symbol of hydraulic mission. This was all the more so for Lesotho. Prior to the implementation of the LHWP, there was little major hydraulic infrastructure in Lesotho. The Basotho hydraulic mission effectively started with the joint feasibility studies, agreement and implementation of the LHWP. The LHWP presented Lesotho with a chance to increase its hydraulic control over the headwaters of the Senqu River. A project of this scale would have been impossible for Lesotho to fund and implement unilaterally. It should be noted that the preliminary agreement was ratified two years after Lesotho's independence, at a time when the government was seeking ways to increase the country's pace of development (see Mirumachi 2008: 40). The mid-plan review of Lesotho's second Five-Year Plan (1975–1980) stated: 'There is a realistic hope that efforts in the natural resources sector can provide employment which will be both competitive with opportunities across the border and reduce imports' (Central Planning and Development Office 1977: WM-1). Governmental policy was to support the use of the Senqu River waters for economic development, and the project was seen as a way to replace export revenue – lost as a result of the declining mining industry (Boadu 1991: 700).

For the South African hydrocracy, it was imperative that the Orange–Senqu River was controlled and tamed to suit society's needs. Titles of project documents, such as *Taming a River Giant* (Department of Information 1971), illustrates this geographical imagination of a basin requiring human intervention. The Orange River Project, proposed in 1962, was symbolic of the South African hydraulic mission with extensive water transfer tunnels and large dams. The project objective was to supply water to existing and newly irrigated areas, as well as to generate hydropower (Department of Information 1971). Increasing urbanization resulted in increased water demand, and the Vaal Dam had doubled its capacity to 2,188 million m^3 by the early 1950s (DWAF 2014a). During the 1960s and 1970s, the construction of large-scale infrastructure in the mid- and lower Orange–Senqu River basin, such as the Vanderkloof and Gariep dams, was undertaken. The Orange River Project was justified on the grounds that 'water will be the greatest limiting factor in the development of the Republic' and that hydraulic development would achieve 'the principle of the highest possible yield per unit of water' (Secretary for Water Affairs 1971: 5). This extent of South African influence upon the use of water resources in this basin is enabled, not only by geography but also by political circumstances. Put differently, South Africa has the largest basin area of the four riparian states and has been able to establish sovereign claims to its water and territory. The Molopo River in Botswana is ephemeral and does not contribute to the water flow in the basin (Heyns 1995: 479). Namibia only gained independence from South Africa in 1990, and its history of river development in the Lower Orange sub-basin is comparatively short. The Orange–Senqu River basin was developed, in large part, in accordance with South Africa's hydraulic mission.

A key feature of the hydraulic mission of the South African hydrocracy is water diversion through tunnels and canals. Diverting the river flow was a prominent solution to mitigating water scarcity as early as 1928, when droughts in the Eastern Cape prompted the Department of Irrigation to investigate ways to secure water. The Orange River Project stems from such ideas evolved through policy papers such as the first White Paper on water resources development in 1951. Within this grand design of the Orange River Project, the LHWP was just one part of extensive river engineering. The fact that the LHWP, which also diverts river flow, was embedded within this larger engineering project is significant because it directly tied domestic water development to international water development. The advanced stage of the hydraulic development in South Africa enabled the hydrocracy to have more resources and expertise to store, divert and transfer water than its Basotho counterpart, strengthening the exploitation capacity needed for hydro-hegemony (see Zeitoun and Warner 2006).

Leaving a mark on the geography of the river basin exercises South African hydro-hegemony. Such hydro-hegemony is further strengthened by its foreign policy influence, best evidenced in the securitization of the LHWP. This securitization was one form of attaining the consent of Lesotho for geopolitical interests. At a time when there was increasing opposition to the apartheid regime from its neighbours, the South African government implemented 'railway diplomacy', offering 'rewards' such as the construction of tracks, ports and roads in return for the termination of military attacks (Jaster 1988: 132–135).

Domestic water politics and hydro-hegemony

Challenging the hydro-hegemonic control over water resources development and planning has been difficult for the Lesotho hydrocracy. Some bargaining power has been exercised by Lesotho. For example, it took advantage of infrastructure expansion by including the Muela Hydropower Plant in the LHWP plans. This 72 MW hydropower plant was a 'strategic investment' to reduce the demand for importing electricity from South Africa (LHDA n.d.b: 15). In addition, during the implementation of Phase 1, the Basotho hydrocracy was able to call upon the authority of the World Bank, one of the project lenders, to ensure that the South African government would meet its obligations to address the socio-economic impacts of the project. The World Bank set up an independent Panel of Experts (PoE) for the LHWP to fulfil its responsibility as a lender to conduct appropriate environmental and social assessment of the project. In 1991, the PoE disapproved of the JPTC's slow progress in achieving rural development initiatives. The LHDA drew upon the importance of the recommendations made by the PoE to sustain pressure about these initiatives (former PoE member 2010, pers. comm.).

However, the negotiations over Phase 2 further exemplify the power asymmetries between the two basin states, in that the South African hydrocracy was able to actively stall commitment to the next phase. Zeitoun (2006: 244–245) argued that active stalling enables the hydro-hegemon to maintain a skewed water allocation arrangement, which benefited its water demands. For example, in the Nile River basin, Egypt exercised its hydro-hegemony by participating in cooperative initiatives, but stalled on their implementation. This has enabled Egypt – a downstream state – to maintain advantageous water allocation, at the expense of the upper basin states (Cascão 2009b: 174–175). Independently investigating alternative plans and the lengthy process of formalizing agreement to Phase 2 were the ways in which the decision to commit to Phase 2 was stalled. The feasibility study of the TWP, the alternative project in KwaZulu-Natal, provided scientific evidence to the technical possibility and economic rationale (DWAF 2000, 2001). The TWP had the added benefit of providing opportunities of poverty reduction for local communities in KwaZulu-Natal (DWAF 2014b). Water transfer projects need to be robustly justified, once the National Water Act introduced in 1998 made fundamental changes to the way in which water is managed and allocated within South Africa. In an effort to redress the inequity in water access and allocation created by the apartheid structure, the National Water Act encouraged inclusive decision-making (Mirumachi and van Wyk 2010). In addition, it also emphasized the importance of water for the environment, thus bringing about new measures for water conservation and recycling. Water transfer projects have to demonstrate its sustainability and utility.[3] While not completely denied a role, implementing inter-basin water transfer projects is more limited compared to previous phases of water resources management.

These domestic concerns contributed to active stalling, which had the effect of determining the pace of bilateral hydraulic development suited for South Africa. After the initial agreement in 2008 to proceed with the project, feasibility studies for Phase 2 were completed in 2011, leading to a formal agreement later that year. When the Minister of Water Affairs and Forestry,

Lindiwe Hendricks, announced the agreement of Phase 2, she carefully phrased the reasons for investing in a project that would cost over R7 million (Hendricks 2008). First, the project was considered as a 'strategic intervention' to alleviate water shortage. The project would be operated along with water demand management strategies such as water conservation, water quality improvement and the reduction of illegal water extraction. Second, the project is described as being energy-efficient:

> The motivation for selecting the Lesotho Highlands Water Project Phase 2 as the preferred option for the augmentation of the Vaal include: the project has a low energy requirement in that water can be transferred under gravity to South Africa without pumping – unlike the Tugela [*sic*] option, which is energy intensive as water must be pumped from the Thukela [*sic*] River over the escarpment. Furthermore the existing hydro-power generation capacity of the Lesotho Highlands Water Project Phase 1 can also be increased. The project would bring substantial benefits to Lesotho as well as a regional benefit, as it will mean the prevention of increased carbon emissions.
>
> (Hendricks 2008: section 3)

Compared to the way in which Phase 1 was described, there was less emphasis on water transfer as being an optimal method for increased water supply. None the less, the next phase is justified to be in line with the ideals of the new water policy. Moreover, a new dimension to the framing of the LHWP may be seen: a means of green energy supply. The project is couched in the context of water *and* energy, highlighting the project's sustainability credentials.

The speech act of committing to the second phase shows that the South African hydrocracy was still implementing large-scale projects but the changing water policy context made it difficult to justify engineering solutions purely for their economic benefit. Here we can see the hydraulic mission overlapping with another water management paradigm, described as reflexive modernity by Allan (2002). This paradigm encourages efficiency of water use through economic means, taking into consideration lessons learned on impacts to the environment from the hydraulic mission. At the same time, it requires a socio-political change to make feasible environmentally conscious policies on conservation and stewardship.[4] The overlap between the two paradigms results in a discrepancy between the new water policy, which emphasizes reflexive modernity, and actual water management practices. This is a good example of how domestic water politics influence decisions and water management practices at the international level. Hydro-hegemony is made up of scalar politics consisting of the development of national water policies, and inter-state strategies and tactics.

Box 5.1 Operation Boleas: Securitization of the Katse Dam?

In September 1998, a military intervention organized by the Southern African Development Community (SADC) took place in Lesotho. Operation Boleas resulted in casualties at the Katse Dam site, a key infrastructure of the LHWP. The South African Defence Force and the Botswana Defence Force were mobilized to contain instability within Lesotho following general elections, after the Basotho Prime Minister requested support from the SADC in writing (see Ambrose 1998). Before launching the military intervention, the South African government announced that it aimed 'to create a safe environment by securing or controlling the Maseru Bridge Border post, the Lesotho Defence Force Military Bases, the Radio Broadcasting Station, Embassies and SA High Commission, Royal Palace, Airports, Government Buildings, *power and water supply facilities*' (Government of South Africa 1998: n.p.; emphasis added). There is analysis claiming that this event is a case where South Africa used its superior military powers to secure water (Pherudi 2003; Likoti 2007) and that it is a case of the securitization of water resources (Davidsen 2006).

The existence of the Katse Dam was significant in this serious military action but it is unclear whether the securitization was solely over water issues. As Davidsen (2006: 52) acknowledged, analysing the actual incentives of the South African Defence Force in intervening in Lesotho is complex and requires concrete evidence. It is unclear whether the securitizing move was based on South Africa's construction of a threat related to water resources management to securitize the issue of water transfer. Rather, a plausible explanation is the tension between the political parties, the Lesotho Defence Force and the government becoming a threat to national security such that external intervention was necessary. Operation Boleas is thus better explained as part of violent conflict that arose from political instability, rather than over water resources. See Mirumachi (2008: 50–52) for details of the Basotho domestic political background to Operation Boleas.

The LHWP and basin-wide water governance

While the LHWP is a bilateral project in the upper basin, this project has interesting implications for the governance of the Orange–Senqu River basin as a whole. The Orange–Senqu River Commission (ORASECOM) was set up as a RBO serving the four basin states. This RBO, in turn, comprises part of a larger governance scheme of the Southern African Development Community (SADC). For SADC, the governance of transboundary river basins is no trivial issue. Within the geographical boundaries of its 15 member states, there are 21 international transboundary river basins shared by 12 states (Turton 2010: 8). This high concentration of international transboundary river basins has given rise to agreements, which over time have emphasized water sharing and environmental stewardship (Lautze and Giordano 2007). The *Protocol on Shared Watercourses Systems in the Southern African Development Community Region*, drawn up in 1995 and revised in 2000, called upon several global policies and legal frameworks. The original SADC Protocol emphasized principles such as equitable and reasonable utilization, environmental considerations, and information and data sharing. These principles refer to international frameworks for water governance such as the *The Helsinki Rules on the Uses of the Waters of International Rivers* adopted by the International Law Association and those on sustainability such as the *Dublin Statement on Water and Sustainable Development* and *Agenda 21* (Ramoeli 2002: 106–107). Among the multiple water institutions within SADC, those relating to the Orange–Senqu River basin are considered to be particularly well developed (Jacobs 2012).

Academic literature argues that cooperation over water may be better understood by examining the types of norms required, and the nature of their diffusion. Notably, Conca (2006) discussed how different norms relating to cooperation over water (what he termed the principled content of agreements) are negotiated, taken up and rejected at different spatial scales and between multiple stakeholders. Taking a cue from this study, Jacobs (2012) argued that norm diffusion has indeed occurred in the Orange–Senqu River basin, evidenced by the presence of global norms that influence regional practice. For example, the Revised SADC Protocol was informed by a global legal framework, the United Nations Convention on the Law of Non-Navigational Uses of International Watercourses (hereinafter UNWC). Even if only South Africa and Namibia have formally ratified the UNWC, the SADC protocol exemplifies that the global water law principles regulate state behaviour. National policies also reflect principles embedded either in the UNWC or regional SADC water policies (ibid.).

In theory, the LHWP would thus exemplify the practice of equitable and reasonable use of shared waters, operating the spirit of cooperation laid out in the ORASECOM mandate and the Revised SADC Protocol. However, the reality is much more complex. The ORASECOM was established in 2000 to achieve sustainable development and with the aim of enhancing cooperation among the four basin states (ORASECOM 2000). This was the first basin-wide technical commission within the basin states, and also the first of its kind within the SADC Protocol (Turton and Funke 2008: 59). This RBO has the broad mandate to offer technical advice to the basin states relating to water use and conservation, disaster management, assessment of development plans, information and data collection and sharing, and dispute resolution (ORASECOM 2000). However, as Article 1(3) of the ORASECOM agreement specifies, it does not change the rights and obligations of the basin states which pre-date its establishment. This clause poses limits on the extent to which ORASECOM can influence the LHWP. Because the LHWP was already in place when the RBO was established, the rights and obligations of Lesotho and South Africa (laid out in the 1986 treaty) remain unaffected. Consequently, ORASECOM can only advise on issues of *future* basin-wide allocation, based on existing arrangements for water transfer set out in the LHWP.

The ORASECOM is tasked 'to strike the balance with a certain status quo already established', and finds itself in a very challenging position (Tekateka 2011: 258). It is interesting to note that the ORASECOM was initiated by Namibia originally as the Joint Permanent Orange River Basin Commission (Heyns 2003). The government of Namibia sought to increase influence having been largely excluded from key decisions in the first phase of the LHWP, as it was not an independent state at the time. It is reported that a '"no objection" statement' was requested from the Namibian authorities, as a way to green-light the LHWP construction. But this clearance was only applicable to the first phase; therefore Phase 2 required renewed consideration of downstream water needs (Heyns 1995: 489). Decision-makers of Namibia were particularly concerned with the level of water availability downstream in a very engineered river basin. This concern was exacerbated by the lack of quantitative indicators for water allocation in both the bilateral PWC and multilateral ORASECOM institutions (Kistin *et al.* 2009: 12).

The ORASECOM has not proved to be an effective platform as the decision-makers in Namibia had hoped. This may be attributed not only to the rules relating to previous arrangements but also to hydro-hegemony being exercised. The South African hydrocracy has maintained that ORASECOM meetings are only for the discussion of multilateral issues. Consequently, specifics of bilateral Phase 2 negotiations were not on the agenda (Kistin Keller 2012: 46). This directive speech act had the effect of limiting discursive space to consider implications of a bilateral project at a multilateral forum. This is an example of sanctioned discourse, a hydro-hegemonic tactic to ensure no change to water allocation (Zeitoun and Warner 2006). Seeking alternative venues for discussion of Phase 2 issues, the Namibian hydrocracy requested observer status in the bilateral negotiations between Lesotho and South Africa. However, this request was strongly rejected by the upper basin states, their reasons including the importance of bilateral discussions to be efficient, and avoiding unnecessary burden and cost of trilateral meetings (Kistin Keller 2012: 46). At multilateral fora, the South African and Basotho hydrocracies operated as a bloc vis-à-vis their downstream neighbour.

Identifying regional benefits

The elite decision-makers of Lesotho and South Africa have couched the LHWP in terms of regional benefit. The Revised SADC Protocol fits this framing, as it considers water to be an input for socio-economic development. This use of a public idea of river basin development for regional benefit is articulated in various assertive speech acts. As a representative of a key agency of the South African hydrocracy, the Minister of Water Affairs and Forestry released speeches that emphasized how sharing water brought about regional cooperation (Hendricks 2006). The speech by the South African President, Thabo Mbeki, when inaugurating Phase 1B is underpinned by the discourse on multilateral benefits:

This project sparkles like a jewel in the crown of the Southern African Development Community (SADC) and the African Union, proving that we can, as Africans, accomplish sustainable development, to the mutual benefit of neighbouring countries and as an example of projects that are needed all over our continent to achieve our renaissance.

(Mbeki 2004: n.p.)

From the South African perspective, the LHWP is a means to achieve the vision of 'water form[ing] a network for economic integration' that is institutionally supported by SADC (South African governmental official A 2006, pers. comm.). Similarly, from the Basotho perspective, the LHWP offers lessons for other bilateral/multilateral water projects in the region (Basotho governmental official 2006, pers. comm.).

However, this regional benefit obscures and diverts attention away from the downstream concerns that the Namibian hydrocracy has raised. Moreover, it once again presents benefits defined from those within the hydrocracy and elite decision-making circles, not necessarily from the perspective of individuals and livelihoods that rely on water resources. While the basin-wide governance structures are important, principles embedded in policies such as the Revised SADC Protocol homogenize the state as an entity with stakes in water resources. The rights and responsibilities of the state are framed in a way that makes it difficult for local communities to be an agent that shapes the benefits, trade-offs and rights that could be sought through a regional framework.

It is true that public participation features in a major way in the policies of SADC. The Revised SADC Protocol emphasizes public participation and identifies Integrated Water Resources Management (IWRM) through which it should be coordinated. IWRM is a framework to jointly consider water and other natural resources such as land from the perspective of equity, efficiency and consideration of the environment. A main feature of this framework is public participation for better decision-making (GWP 2000). Regional policies, such as the Regional Strategic Action Plan, refer to IWRM (see ORASECOM 2014). ORASECOM has made efforts to establish best practices of public participation, noting the challenges of including and maintaining the engagement of a range of actors in a large basin area (SADC and EDF 2009). At the national level, basin states have updated policy and legislation regarding water resources management. As mentioned above, South Africa's National Water Act is progressive, emphasizing cooperative practices among different stakeholders (Mirumachi and van Wyk 2010). The Water Act of Lesotho established in 1998 explicitly subscribes to the principle of public participation. Namibia's policies, namely the Water Resources Management Act of 2004 and the accompanying Water Resources Management Bill approved in 2010, also encourage a wider stakeholder group for decentralized management and planning, and the Department of Water Affairs of Botswana has recently devised strategies to manage and plan water resources development based on IWRM principles. The strategies highlight the range of stakeholders required for inclusive decision-making (DWA 2013).

Public participation may also be facilitated through the various water governance initiatives at multiple scales. It is not only SADC or ORASECOM that can provide discursive space. In the lower Orange–Senqu River basin, issues of water use are discussed between South Africa and Namibia through the Permanent Water Commission established in 1992. Project-based discussions on water use are also conducted through the Joint Irrigation Authority, also set up in 1992. This organization is responsible for the Vioolsdrift and Noordoewer Joint Irrigation Scheme. However, it has been pointed out that the capacity to support decision-making is needed not only at the national scale, but also at the local scale in the Orange–Senqu River basin (Mokorosi and van der Zaag 2007). Problematically, public participation at the project level has been limited in the case of LHWP (Hitchcock 2012; see also Haas *et al.* 2010). This makes bridging public participation across scales difficult. In addition, it may be questioned whether 'regional' benefit can even be defined in a way that not only addresses power relations between states but also within states, if not to empower the majority excluded from elite decision-making.

Challenging hydro–hegemonic control?

This chapter has shown how bilateral transboundary water interactions concerning the LHWP between Lesotho and South Africa have changed over time. One of the main factors responsible for this change is the broader political context in which these interactions occurred. The TWINS framework enables a comprehensive way to understand the process of transboundary water interaction against the political backdrop of this region. Thus, the analysis presented provides a more critical view of how the transboundary water interaction has evolved, unlike analysis that examines only one dimension of the project or snapshot of relations as 'a good example of upstream/downstream cooperation and benefit-sharing over an international watercourse' (Baillat 2010: 94). Issues of geopolitics and apartheid diplomacy were entangled with water transfer negotiations. This example of Lesotho and South Africa indicated how low politics is not always contingent upon high politics. In fact, the case study demonstrated that high politics and low politics do mix. To this end, the concept of securitization is useful. It exemplifies the way in which environmental issues can be prioritized on bilateral agendas to the extent that no alternatives are considered, bar those suggested by the hydro-hegemon. In the case of the LHWP, high and low politics are not so clear cut and delineated, contrary to existing analyses such as that by Abukhater (2013: 185). The differentiation between high and low politics is too simplistic, and the implications of mixed high and low politics yield some interesting insights. As the previous example of the Ganges River basin (Chapter 4, this volume) showed, the securitizing practice of 'cooperation' has significant implications for the allocation and use of water resources. In this Southern African case, the treaty of the LHWP has set a status quo of water allocation that downstream Namibia has found hard to challenge, even though basin-wide governance emphasizes cooperation.

Study of this river basin shows that cooperation is not without its risks. This is especially true for Namibia, having proposed an alternative forum to discuss its concerns but resulting in a basin-wide RBO that consolidates upstream hydro-hegemonic control. Multi-scalar water governance does not seem to radically alter such control. Within ORASECOM there is no guarantee that contentious issues of water allocation will be solved, or even addressed. Hydro-hegemonic control by South Africa is evident. Kistin Keller (2012: 52) pointed out that the bargaining power exercised by South Africa has been strengthened by Lesotho being 'a significant ally ... in the perpetration of the bilateral planning model.... Botswana [is] unlikely to push for more substantive engagement at the basin scale'. Certainly, the number of cooperative institutions present at various levels is not a useful indicator of the nature of transboundary water interaction.

The geography of the river basin has been marked by domestic hydraulic mission projects. The attempts to change the river flow underscore the interests of the elite decision-makers to make the basin an economically productive one. The geographic imagination of the river expresses governmental intervention as necessary and instrumental for society's well-being. While there is rhetoric of regional benefits to be gained from developing the river, the way the river flow is engineered and modified emphasizes *national* water planning – and ultimately the sovereignty over its share of water. The Orange–Senqu River basin demonstrates a very territorialized space.

Notes

1 Formerly the Joint Permanent Technical Committee (JPTC).
2 The LHWP has also been controversial because of a corruption case between the Chief Executive of LHDA and a group of construction companies and consultancies (see Earle 2007).
3 The *White Paper on a National Water Policy for South Africa* stated:

> Inter-basin transfers will have to meet special planning requirements and implementation procedures, which must involve agencies from both the donor and recipient catchments. Catchments to which water will be transferred will have to show that the water currently available in that catchment is being optimally used and that reasonable measures to conserve water are in force.
>
> (DWAF 1997: section 6.6.3)

4 See Gilmont (2014) for an example of the challenges of this paradigm to moderate water consumption, or 'decoupling'.

6 Developing the Mekong waters

Introduction

Cooperation over water resources within the Mekong River basin is often associated with the 'Mekong spirit', or the goodwill of the states to work together despite being political adversaries. It is said that the Mekong spirit has carried the discussion forward on river basin development without getting bogged down by political differences (Menon 1972: 168; Jacobs 2002: 358). Since the 1950s, the four lower basin states – Laos, Thailand, Cambodia and Vietnam – have deliberated the ways in which the shared river may be used to develop navigation, fisheries, irrigation and hydropower. On the whole, more of the water use has been in the lower basin than in the upper areas, and there have been comparatively active investigations, planning and execution of water resources development between the four states. Large-scale consumptive water use on the mainstream in upstream China and Myanmar has hitherto been limited by the steep topography; moreover, its overall contribution to the annual mean average flow is 18 per cent in comparison downstream (MRC 2005b: 7; MRC 2010: 10). For the four lower basin states, river basin development must contend with controlling the river flow: managing high river flow from monsoon rainfall *and* problems of water shortage during the low-flow season (Le-Huu and Nguyen-Duc 2003: 3).

The development initiatives for better agriculture and livelihoods, energy production and navigation attempt to take advantage of the 760,000 km^2 large basin, endowed with a high annual runoff of approximately 457,000 million m^3 (MRC 2010: 20) (see Figure 6.1). However, using the water resources of this shared river has caused tension among the lower basin states. Furthermore, with Chinese dam development now progressing at a rapid pace, contention between China and the lower basin states is increasingly prominent. More generally, large-scale development projects are heavily scrutinized and questioned by international environmental activists and regional civil society groups. The current RBO of the lower Mekong basin, the Mekong River Commission (MRC), is being challenged to provide a collective response to these emerging environmental concerns and contestations.

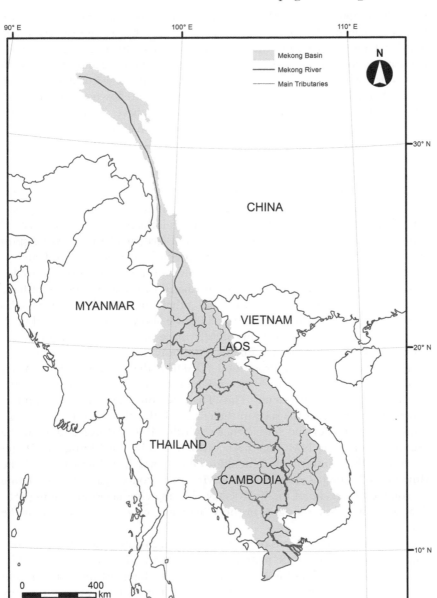

Figure 6.1 Map of the Mekong River basin.

The unique institutional development of the Lower Mekong River basin serves to probe the question: how do transboundary water interactions manifest in RBO decision-making? Stemming from the purpose of this book to understand how and why coexisting conflict and cooperation occur among states, this chapter explores the ways in which norms and activities of RBOs have been shaped, and continue to be shaped, since they were first established in 1957. Thus, compared to the previous two cases, attention is given to the negotiations concerning institutional development, rather than to projects per se. The following analysis of the Lower Mekong River basin is focused on Thailand and Vietnam, in order to illustrate upstream–downstream dynamics between two rival regional powers. Both states have been continuously involved in three RBOs, and instrumental in negotiations within this arena. The institutional history of the Lower Mekong River basin has been well documented and discussed (e.g. Sangchai 1967; Takahashi 1974; Jacobs 1994, 1995, 1999; Hori 1996, 2000; Chi 1997; Makim 1997, 2002; Browder 1998; Browder and Ortolano 2000; Öjendal 2000; Menniken 2008). The originality of the analysis presented in this chapter is in the constructivist interpretation of riparian *interaction* that influences and is influenced by the multilateral institutions. By doing so, it provides a critical analysis of the Mekong spirit and the governance implications for contemporary issues of water and energy.

This case study portrays high stakes involved in the negotiations of multilateral basin institutions. This process is mediated by the hydrocracy that has a vested interest in river engineering and other elite decision-makers who support the hydraulic mission (Wester 2008; see also Chapter 3, this volume). The Thai hydrocracy has been historically made up of an array of organizations, including the National Energy Authority (later the Department of Energy Development and Promotion (DEDP)), the Department of Water Resources in the Ministry of National Resources and Environment, and the Royal Irrigation Department. The Thai National Mekong Committee (NMC) that represents the state in the current RBO thus coordinates a large number of ministries and line agencies (Griffin 1991; MRC 2007). In the case of Vietnam, the NMC has been variously led by the Ministry of Water Resources, the Ministry of Agriculture and Rural Development, and the Ministry of Natural Resources and Environment, once again demonstrating the complex make-up of the hydrocracy. Tracing elite decision-making provides some interesting insights on how 'national' views are presented at the RBO.

Understanding speech acts of the two hydrocracies and associated decision-makers using the TWINS matrix (see Chapter 3, this volume) provides inroads to understanding the heightened dam development interest in recent years. I argue that hydro-hegemony pervades deliberation on water for energy through the historical institutionalization of river basin development. While geopolitical factors are important to understanding the context of such institutionalization, regional and domestic political economies for water and energy also play an important role. This case study points to an important insight into how constructed scales of water resources management influence dam and energy debates.

The Mekong Committee and river basin planning

The development of the lower Mekong River basin came to the attention of Laos, Thailand, Cambodia and South Vietnam through a report by the United Nations Economic Commission for Asia and the Far East (ECAFE) in 1952.[1] This report, *Preliminary Report on Technical Problems Relating to Flood Control and Water Resources Development of the Mekong – An International River*, suggested potential development plans that included a series of mainstream dams. Inspired by the Tennessee Valley Authority in the USA, it was envisaged that major hydraulic engineering could assist irrigation and provide for hydropower generation, greatly benefiting the lower basin region (Hori 1996: 86–88). These findings led the four countries to ask the USA for a survey of the lower river basin in 1954. This joint request represents a directive speech act on the part of both the Thai and South Vietnamese governments. This speech act identified the importance of lower basin development and thus politicized river basin planning on the international agenda. The joint request represented transboundary water interaction with a cooperation intensity of ad hoc joint action: while there is joint action, no specific shared goals on water use and river development have yet been developed (TWINS Sequence 1 in Figure 6.2). This development in transboundary water interaction is significant. While there had been several other bilateral and multilateral agreements concerning the Mekong waters since the early 1900s, these earlier agreements focused mainly on navigation issues between the basin states and the colonial powers, rather than on water allocation or development issues (see Chi 1997: 194–219).

The interest in the hydraulic engineering of the river culminated in three scientific studies that put forth large-scale dam development on the mainstream. In addition to the 1952 ECAFE report, the *ICA Reconnaissance Report on the Lower Mekong River Basin* produced by the US Bureau of Reclamation in 1956, and the *Development of Water Resources in the Lower Mekong Basin* in 1957 by ECAFE as a follow-up to their 1952 study, all pointed to the wealth of opportunities to use the water resources for economic development. For this, it was emphasized that close riparian cooperation would be necessary to implement large-scale projects (Hori 1996: 97–98). To continue investigations into river basin development, the four states established the Committee for the Coordination of Investigations in the Lower Mekong Basin, commonly referred to as the Mekong Committee (MC), with the support of ECAFE in 1957.[2] The original idea conceived by ECAFE to develop the Mekong River for socio-economic progress was embraced by the four states and institutionalized (Sewell and White 1966: 49).

The MC was guided by the *Statute of the Committee for Co-ordination of Investigations of the Lower Mekong Basin* (hereinafter Statute of MC). In this Statute, the role of the MC was to 'promote, coordinate, supervise and control the planning and investigation of water resources development projects in the Lower Mekong basin' (Statute of MC 1957: Article 4). The

Statute established the common norm of basin-wide water resources planning and development, with the goal of constructing hydraulic infrastructure, particularly for the use of the mainstream waters. The establishment of the MC and the norms guiding the activities of the organization is a commissive speech act indicating the intent to achieve regional water resources development. Joint action to achieve this goal was reflected in the efforts to collect and assess data for the planning and implementation of hydraulic projects (TWINS Sequence 2 in Figure 6.2). It should be pointed out that China and Myanmar (Burma) were not involved in the MC because the former was not part of the United Nations community at the time, and the latter was not interested in participating (Mekong Secretariat 1989: 10–11).[3]

River engineering planned by the MC was extensive. Seven mainstream dams were envisaged, using the drop in elevation as the river flowed to the delta. Situated along the mainstream were the High Luang Prabang, Sayabouri, Pa Mong, Upper Thakhek, Ban Koum, Stung Treng and Sambor dams. These dams comprised the Mekong Cascade, designed to control floods, generate hydropower and contribute to the expansion of irrigation. In addition, there were plans to manage water flow into the Tonle Sap, a flood pulse lake in Cambodia, and to prevent saline intrusion in the delta areas of South Vietnam (MC 1970). The Indicative Basin Plan Report published by the MC in 1970 summarized the extensive choices and options to develop the river. A series of reports conducted with the aid of external organizations added authority to the grand design of the river: *Programme of Studies and Investigations for Comprehensive Development of the Lower Mekong River Basin*

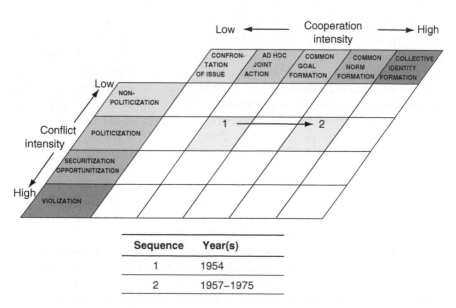

Sequence	Year(s)
1	1954
2	1957–1975

Figure 6.2 TWINS matrix of Thailand–Vietnam relationship: Sequences 1–2.

(1958) by the Wheeler Mission under the US Army Corps of Engineers; *Comprehensive Reconnaissance Report of Lower Mekong Basin* (1961) by the Government of Japan; and *Economic and Social Aspects of Lower Mekong Development* (1962) by the Ford Foundation Mission, led by Gilbert White, a geography professor from the University of Colorado. As a means to bring these plans to reality, the Mekong Secretariat, which supported the administrative duties of the MC, prepared the *Lower Mekong Basin Water Charter* in 1971. The aim was to define the rights and duties of the basin states regarding water resource use. Eventually, this extensive document was revised and summarized in the *Draft Statement of Principles for the Planning and Utilization of the Waters of the Lower Mekong Basin* (Nakayama 1999: 296–297). A geographic imagination of the river ready for development gradually gained materiality through reports, mandate documents and the mobilization of resources for the MC.

Despite this geographic imagination, it was this process of rule-making over water use that became highly politicized, particularly between the Thai and Vietnamese hydrocracies. Both actors were interested in developing waters. The Thai hydrocracy was particularly keen to develop the northeast region within the country where the Mekong waters could benefit irrigation expansion and promote economic development for a not insignificant portion of the population (Mirumachi 2012: 93). The Vietnamese hydrocracy was broadly supportive of river development initiatives because regulating river flow through dams would facilitate increased dry-season flow, thus enhancing agricultural efficiency in the delta area (Mekong Secretariat 1989: 9). However, the Vietnamese representative to the MC argued that the Draft Statement should have the power to monitor and enforce rules in order to ensure downstream water flow and quality (Nakayama 1999: 297). In contrast, the Thai hydrocracy made a directive speech act by emphasizing that 'agreement in the form of the Water Charter ... "as a mere declaration of principles [of water utilization] along with the line of the Declaration of Human Rights"' (ibid.: 297, citing Mekong Committee 1974: 5). From the Thai perspective, the document would simply be a guideline for conduct. This development in the bilateral relationship may be interpreted as Thai hydro-hegemony influencing the outcome. Ultimately, the *Joint Declaration of Principles for Utilization of Waters of the Lower Mekong Basin* (hereinafter Joint Declaration) was eventually established without any binding effect on the basin states (ibid.).

This major institutional development distinguished two important norms over water use. First, mainstream waters were treated as 'a resource of common interest' to the four basin states (Joint Declaration 1975: Article 10). This meant that any development project using the mainstream waters had to gain unanimous agreement and compile a formal agreement signed by all four states (ibid.: Article 17). This rule, however, meant that states can effectively abstain from agreeing to a project or use their 'veto powers'. Second, ensuring equity among upstream and downstream states was reflected in maintaining a certain level of minimum flow during the dry season (ibid.: Article 18).

The discussions over water use among the four lower basin states marked a new era of river basin development where basin-wide plans were investigated. At the same time, these deliberations over norms and guiding principles reflect the upstream–downstream contention over water allocation.

The Interim Mekong Committee and regional instability

When the 1975 Joint Declaration was established, it was hailed as 'a cogent expression of the spirit of the Mekong Committee and a fine example of a highly constructive attitude in regional co-operation' (MC 1975: 19). However, the reality of achieving regional cooperation was constrained by major geopolitical changes. The progression of the Second Indochina War heightened political instability and intensified rivalry between states. The region became a geopolitical hotspot of competition between US and communist influences. When that war ended in 1975, a unified, communist regime in Vietnam was established. Communist governments were also established in Laos and Cambodia during the same year. In this unstable climate, the MC carried on some activities relating to flood forecasting and agriculture, but concentrated largely on research, including revising the feasibility of plans laid out in the 1970 Indicative Basin Plan that identified the Mekong Cascade, the consideration of alternative plans and data collection (Jacobs 1995: 143; Hori 1996: 164). Ultimately, the ideological split between Thailand and the rest of the lower basin states resulted in a situation where Vietnam, Laos and Cambodia failed to send representatives to the MC between 1976 and 1977 (Mekong Secretariat 1989: 46). The Mekong Secretariat existed but without any riparian representation; basin-wide development was not possible. This lack of joint activity on hydraulic development reflects reduced cooperation intensity in the bilateral relationship between Thailand and Vietnam (TWINS Sequence 3 in Figure 6.3).

In an attempt to ensure cooperation, ESCAP facilitated informal meetings among Thailand, Vietnam and Laos in 1977 without the participation of Cambodia, which, under the Khmer Rouge regime, isolated itself from the international community. The UN Under-Secretary General for Inter-Agency Affairs and Coordination recommended an interim committee be established by the three states in order to keep up the momentum of river basin development (Bangkok Post 1977). The involvement of the UN organization resulted in the three states signing a joint communiqué in April 1997, reaffirming their intention to cooperate on shared waters and to establish an interim committee (The Nation 1977). The Thai hydrocracy viewed this multilateral committee as a necessary organization for river basin development for several reasons. First, the MC had been a useful vehicle to generate funds for its own hydraulic development. Second, the MC had facilitated river basin planning and information exchange between the basin states. Third, the Thai hydrocracy recognized the potential for a multilateral committee to facilitate an agreement for mainstream development, which would greatly enhance the economic development of its northeast region (Mekong Secretariat 1977a: 93–96). The Vietnamese hydrocracy was not opposed to the idea of the interim committee. This was because it expected that the 1975 Joint Declaration would continue to safeguard downstream water use by preventing water pollution and salinity intrusion from the low water flow, and ensuring water regulation by upstream storage dams (Mekong Secretariat 1977b: 20–22).

The three lower basin states established the Interim Mekong Committee (IMC) in 1978. The re-establishment of the multilateral committee may be seen as a commissive speech act that (re)formed goals 'to reap the benefits of the development of the water resources of the lower Mekong Basin, to meet the needs for reconstruction and economic development of their respective countries' (Declaration 1978: Article 1). The *Declaration Concerning the Interim Committee for Coordination of Investigation of the Lower Mekong Basin* maintained the common norm of basin-wide water resources development (TWINS Sequence 4 in Figure 6.3). The lack of Cambodian representation meant that no mainstream projects could be developed because obtaining full quadrilateral consent would be impossible. None the less, the interest in the Pa Mong dam, one of the mainstream projects of the Mekong Cascade, was sustained and some new initiatives to better understand and forecast the hydrology of the basin were taken up (Jacobs 1995: 144).

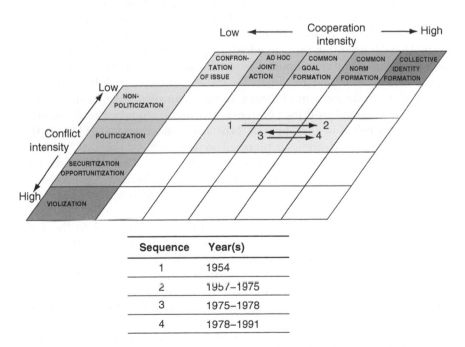

Sequence	Year(s)
1	1954
2	1957–1975
3	1975–1978
4	1978–1991

Figure 6.3 TWINS matrix of Thailand–Vietnam relationship: Sequences 3–4.

Makim (1997: 179) argued that the IMC was important to Thailand and Vietnam in particular, because it acted as a platform to indirectly advance, challenge and restrain the states' political actions. An example of this is when Thailand withdrew from the twenty-first IMC Session in 1985. Political relationships had soured between Thailand and Vietnam, after the latter had invaded Cambodia and maintained a military presence. When an incident involving Vietnamese military intervention near the Thai–Cambodia border resulted in Thai casualties, the Thai government used the IMC as the political stage to register discontent. Military interference was a long-standing and thorny issue between Thailand and Vietnam, and this particular incident sharply brought into focus the rivalries between the two states (Makim 1997: 163; Browder 1998: 60). The Thai government considered this incident to be a serious attack on its sovereignty and was absent from the IMC session in protest (BBC 1985). This directive speech act underscores the politically unstable climate in which transboundary water interaction was taking place. Importantly, it shows that deliberative space created by the RBO was used to further strategic geopolitical interests – an example of Thai hydro-hegemony contributing to regional hegemonic contentions.

Securitizing water use

The end of the Cold War and the gradual resolution of political tension within the Southeast Asian sub-region in the late 1980s provided a new opportunity for the resumption of quadrilateral discussion over basin development. The détente of the US and the Soviet Union relationship served to relax Thai–Vietnam tensions. In addition, the withdrawal of the Vietnamese presence from Cambodia resulted in ameliorated relations between the four lower basin states (Makim 2002: 23–25). Against this backdrop, renewed efforts were made for a quadrilateral arrangement over river basin development. The Cambodian Prince Norodom Sihanouk requested readmission to the IMC on behalf of the Supreme National Council in 1991 (The Nation 1991). A draft declaration was prepared so as to replace the IMC with the MC represented by the four basin states (IMC 1991). This draft declaration served to reactivate the role of the MC as the coordinating organization of water development projects based on the original Statute signed in 1957. Mainstream development would now be subject to the consensus of all lower basin states, based on the 1975 Joint Declaration. Compared to the IMC mandate, these original MC documents contained more detail about decision-making procedures of mainstream and tributary development.

However, the old rules of the MC pertaining to water use and allocation re-emerged as a contentious issue between upstream Thailand and downstream Vietnam. The Thai delegation considered the 1957 Statute and the 1975 Joint Declaration to be too restrictive on water use, reducing opportunities to take advantage of the Mekong waters flowing in its northeastern territory (Makim 1997: 197–198; Browder 1998: 101; Former UNDP official 2008, pers. comm.). The Thai delegation suggested that revisions be made to the documents of the MC and IMC, and that during these negotiations the less restrictive IMC document remain in force. This proposal was not acceptable to Vietnam, Laos and Cambodia, because they intended to use the MC documents from the negotiation period onward (Browder 1998: 101–103).

The issue of water use became securitized with a directive speech act by the Thai Foreign Minister. This securitization of taking extraordinary measures that circumvent usual political practices (see Chapter 3) is observed in the minister cancelling the scheduled Mekong Commission session in February 1992 at the last moment, in an attempt to invalidate the 1957 Statute of MC and the 1975 Joint Declaration (Bangkok Post 1992; BBC 1992; Browder 1998: 104–107). This speech act constructed the former water use rules to be a threat to Thailand's water resources development, and discursively framed the necessity and importance of ensuring options for water resources development. The Thai government effectively took pre-emptive action to avoid losing any options for water development (Former UNDP official 2008, pers. comm.). This speech act was followed by extraordinary measures that were alien to the usual multilateral politics of the RBO. An alternative meeting with states from the entire basin – the four Lower Mekong River basin states, plus China and Myanmar – was proposed. Furthermore, the Thai government expelled the Executive Agent of the Mekong Secretariat, on the grounds that he was biased towards Laos, Cambodia and Vietnam (Handley and Hiebert 1992; Browder 1998: 107–111). The securitizing move undermined the Secretariat and signalled that the governance structure was redundant. By taking these extraordinary measures, the Thai hydrocracy moved the issue of water use out of the realm of normal politics into that of exceptional politics (TWINS Sequence 5 in Figure 6.4).

The securitizing move had the effect of mobilizing external intervention. In 1993, the UNDP began mediating meetings between the basin states. A Mekong Working Group (MWG) was set up to facilitate negotiation and communication among the four states with an aim to develop a new institutional agreement (Radosevich 1996). Securitizing water utilization by the Thai government led to a situation where old institutional agreements were put aside to establish new rules and principles. The *Agreement on the Cooperation for the Sustainable Development of the Mekong River Basin* (hereinafter, Agreement) was signed by the four states in April 1995 and established a new regime in which varying degrees of control over water utilization were clarified. As explained in Chapter 3, and also evidenced in the Ganges case study in Chapter 4, environmental securitization does not necessarily result in overt conflict. Rather, it can lead to measures that attempt to bring about change through agreements and principles to regulate action (Trombetta 2011). This Mekong example provides support for this claim. While the external intervention diverted the issue of water utilization away from that of 'emergency politics' (Roe 2006: 426), a new regime over the development

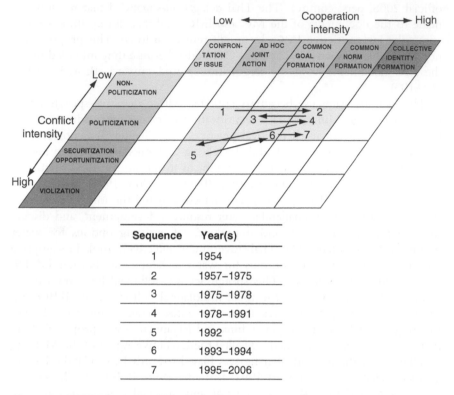

Sequence	Year(s)
1	1954
2	1957–1975
3	1975–1978
4	1978–1991
5	1992
6	1993–1994
7	1995–2006

Figure 6.4 TWINS matrix of Thailand–Vietnam relationship: Sequences 5–7.

and governance of the basin was established (TWINS Sequence 6 in Figure 6.4).

This is not to say that building consensus on the Agreement was without its challenges. Again, Thai representatives attempted to exercise its hydro-hegemonic control over mainstream waters. The Agreement specified in Article 26 determines mainstream water diversion during the wet season requiring prior consultation, while water utilization is less restrictive as long as notification is made to the other basin states. In contrast, dry season mainstream water use required prior consultation that implies the 'aim of arriving at an agreement' as stipulated in international law (Radosevich 1995: 14). No hard rules for implementation were established in the Agreement and instead reflected the very intense upstream-downstream contestation over the degree of freedom in utilizing shared waters.

Browder (1998) provides some useful evidence of the assertive speech acts made during the MWG meetings. The Thai MWG representative, Prathes Sutabar, expressed the desire for cooperation, based on a new institutional mechanism but with a subtle caveat of a 'new appropriate cooperation mechanism' (Browder 1998: 116, citing Mekong Secretariat 1992: Annex VII, p. 1). This speech act implies that water utilization needed to be sought in a way that basin-wide institutions do not interfere with domestic water development plans. But a difference in national perspectives on water utilization and allocation is made by the Vietnamese MWG representative, Phan Si Ky. The representative argued for the 'respect for [...] legitimate interests in the utilizing the Mekong resources'; an implication of its downstream concern that sufficient water flow in the delta area needs to be ensured (ibid, citing Mekong Secretariat 1992: VIII, p. 1). Consequently, the Vietnamese delegation to the multilateral negotiations argued for tighter control of water utilization through prior consultation measures. In contrast, its Thai counterpart proposed and counter-proposed draft article texts to define water utilization in a way that would limit institutional control of water resource projects (Radosevich 1995: 10–14).

The final outcome, in the form of the Agreement, established common norms of sustainable development with regard to shared waters between Laos, Thailand, Cambodia and Vietnam. This norm was underpinned by international law principles, such as reasonable and equitable utilization, and duty to cause no significant harm (Radosevich 1996: 259–260). The signing of the Agreement, and the establishment of the Mekong River Commission, represents commissive speech acts by both Thai and Vietnamese hydrocracies. However, detailed rules to implement water use and allocation in accordance with Article 26 of the Agreement were left outside the bounds of the multilateral negotiations at this stage (TWINS Sequence 7 in Figure 6.4).[4]

The Mekong River Commission and procedures for water use

While the four lower basin states acknowledged the importance of establishing water use and allocation rules, it was not until 2000 that the preparation of such rules began. The Water Utilization Programme (WUP), funded by the Global Environmental Facility, was set up though beset by difficulties in legalizing the rules (former MRC Secretariat official 2008, pers. comm.). An indication of the prevailing degree of politicization is evident in a MRC-commissioned study on the preparation of water quantity rules by the Japan International Cooperation Agency (JICA 2001). Documents related to this study were careful to put the word *rules* in quotation marks, in order to indicate that what was being proposed did not represent a concrete obligation of the basin state. Even so, the nature of the study was deemed so controversial that its key focus was eventually changed to an *assessment of the scientific requirements to understand hydrological flow*. The title of the report, *The Study on Hydro-meteorological Monitoring for Water Quantity Rules in the Mekong River Basin*, reflects this de-emphasis on rule-making (ibid.).

The activities of the WUP were fundamentally frustrated by the long-standing issue around the degree of intervention acceptable in regulating water use. It is, therefore, important to note how the hydro-hegemonic inter-actions among Thailand and its basin neighbours continued to play out within the newly established institution. By signing the Agreement, Thailand accepted a degree of control by the institution over dry season mainstream water use. However, Thai officials claimed that sovereignty should be respected in order to allow free water use within one's own territory (Watershed 2003: 6). A speech by the Vice Minister for Natural Resources and Environment, Prinya Nutalaya, used phrases such as 'thoughtful con-sideration' to subtly reinforce the notion of sovereignty, as below:

> It is quite understandable that there remain some delicate issues to be dis-cussed, which we must try to overcome with thoughtful consideration and on the basis of all necessary data being available to support our con-sideration and judgment.
>
> (MRC 2003a: 49)

This assertive speech act supports a claim made by an interviewee that safeguarding national interests of water use is the hard reality (Former Thai governmental official 2008, pers. comm.). From the Thai hydrocracy perspective, sovereignty was a norm that restricted intervention by the MRC and other states. The final documents prepared by the WUP reflect the rejec-tion of rules. All guidelines are rather entitled 'procedures' in order to avoid any implication of binding legal power being contained in these documents. The *Procedures regarding Notification, Prior Consultation and Agreement* (PNPCA) determined time frames for sufficient notification and prior consultation; necessary information, such as impact assessment and other technical data; and the role of the MRC Joint Committee, including the decision-making powers of the senior riparian representatives (MRC 2003b).

The completion of the WUP in 2006 marks a milestone in both the Thai–Vietnamese relationship and in the institution-building efforts of the four lower basin countries. Five procedures addressed aspects of water use during the wet and dry seasons: *Procedures for Data and Information Exchange and Sharing* (approved 2001); *Procedures for Notification, Prior Consultation and Agreement* (approved 2003); *Procedures for Water Use Monitoring* (approved 2003); *Procedures for Maintenance of Flows on the Mainstream* (approved 2006); and *Procedures for Water Quality* (approved 2011). While water use was still on the public agenda, common norms established by the Agreement were further supplemented by these institutional arrangements. As will be explained below, these arrangements have interesting implications for ongoing dam development and planning on the mainstream. The governance of water *and* energy issues is bound up in the legacy of river basin development and the norms and rules established over time.

Mekong spirit: suppressing conflict, limiting cooperation

Transboundary water interaction between Thailand and Vietnam shaped, and was shaped by, the institutional development of lower basin water resources management. Focus on these two states is useful, as they played significant roles in the development and adoption of norms, principles and 'procedures'. Importantly, the TWINS analysis showed that while the issue of transboundary water has largely been politicized, the level of cooperation intensity has changed over time. The Thai–Vietnamese relationship exemplifies that 'cooperation' under the banner of the Mekong spirit is, in fact, much more nuanced than the existence of a resilient regional water institution since the 1950s. This means that a generalization of the effect of the Mekong spirit is unhelpful; a critical explanation is needed to account for changes to coexisting conflict and cooperation.

It is interesting to note that this phrase, 'Mekong spirit', did not evolve organically from the riparian states. Rather, the basin states adopted it from ECAFE promoting regional cooperation in the early 1960s. U. Nyun, the Executive Secretary of ECAFE, first described the Mekong spirit as follows:

> Much has been achieved by the member countries of the Mekong Committee within a short period of time, but much more still remains to be done. It is indeed very inspiring and encouraging to know that the four riparian countries assisted by the United Nations and the co-operating countries are working together as true partners in progress in the true Mekong spirit of collective responsibility, friendly co-operation and understanding firmly determined to achieve a fuller and richer life for all people who dwell in the Mekong basin.
>
> (MC 1964: annex 1, 6–7)

The Mekong spirit reveals the instrumental creation of a geographical unit by UN organizations and by donor agencies, particularly the US. For the US, this region was of strategic importance in Cold War geopolitics. In order to contain Soviet and communist Chinese influences, the US sought to provide economic assistance to the countries sharing the Mekong River. For the UN agencies, the river basin was a suitable geographical unit to promote international cooperation in the region (Sneddon and Fox 2012). The river itself is associated with the geographical imagination of an unharnessed river, unexploited for economic development, laced with geopolitical intents. This imagination supports the bringing together of the 'Mekong' region (Bakker 1999). The Mekong spirit plays into a discourse which justifies the idea that the Mekong River is unregulated – causing water shortage and droughts – and that its economic potential may be increased through basin-wide hydraulic development, leading to peace and harmony.

However, cooperation based on the Mekong spirit has, ultimately, served the interests of those wishing to postpone and avert decision-making over binding rules for water allocation and river basin development. It is true that both upstream Thailand and downstream Vietnam bought into the idea of basin-wide development for economic growth, as represented in Sequence 2 of the TWINS matrix. None the less, the response to this idea was pragmatic: to aid *national* development projects. During the IMC period, the Thai government constructed pumping stations and hydraulic infrastructure to control flooding (and consequently to enhance irrigation), arguing that investment in such projects through the IMC promoted national economic development (BBC 1986c). For Vietnam, the IMC was an important source of external funding at a time when it was isolated from the international community following its military interference in Cambodia (Makim 1997: 169). National projects, like the Tam Phuong, were constructed in the delta area, with the aim of preventing saline intrusion and improving irrigation. As the TWINS analysis showed, fundamental differences among the upstream and downstream states on the need for mainstream dams and the level of intervention in water use were never fully resolved.

It may be argued that the Mekong spirit also served the interests of the donor community. UN organizations and donor funding were used at critical points of institutional development. For example, the MC Executive Agent, W.J. van der Oord, was instrumental in liaising with Thailand and the communist-led Vietnam and Laos during the institutional limbo in 1977, and secured funding from donor countries for further development activities (Makim 1997, 2002). As the TWINS analysis showed, organizations like the UNEP and GEF were the catalysts for establishing the MRC and giving shape to activities of the basin commission. By promoting 'cooperation' in the lower Mekong basin with the geographical imagination of the river, these external organizations have been able to justify their roles in developing a regional public good.

Water scarcity in a water-rich basin

An examination of the national-level politics of water resources development provides further explanations for the unresolved upstream and downstream contention between Thailand and Vietnam. Looking across scales at which politics play out, or what Mollinga (2008a, 2008b) called the 'domains' of water politics, is helpful in showing how domestic politics determine the interests and positions taken at international negotiations. The scalar politics is made up of narratives of water scarcity, which contributed to the discourse of water use negotiated in the MC, IMC and MRC. As will be shown below, these narratives of water scarcity serve to further emphasize the untapped 'wealth' or 'abundance' of the river that may be exploited through the hydraulic mission.

Vietnam's concerns about maintaining sufficient flow in the Mekong delta may be explained by domestic drivers of water resources development related to agricultural security. Solutions to increased agricultural activity centred on infrastructure solutions, which could help regulate water flow in and out of the delta (Käkönen 2008). Delta development lagged behind, even though rice irrigation capacity gradually increased from the late 1960s to the mid-1970s as a result of rolling out the Green Revolution (Le Coq *et al.* 2004: 175–180). Large-scale projects such as the Mekong Delta Development Programme supported by the MC were considered, but never bore fruit because of the ongoing warfare (Biggs *et al.* 2009: 209). Rice production became an urgent governmental strategy to ease food shortages following the establishment of the communist Vietnamese state in 1975. The government implemented agricultural policies that collectivized rice production, and heavily promoted its 'rice everywhere' campaign (ibid.: 210). Embankments, dykes, canals, sluices and water pumps were constructed to facilitate multiple cropping systems (Käkönen 2008; Barker *et al.* 2010: 269).

According to Tu (2002: 339–342), extensive hydraulic infrastructure construction for irrigation development was funded by 10 per cent of state budget during the 1990s. Seventy-five large- and medium-scale irrigation systems, approximately 600 large and medium reservoirs, and numerous small-scale storage and pumping stations came on line. Consequently, rice irrigation grew to such an extent that by 1994 the size of the irrigated area had doubled compared to that of 1976, and the drained area had increased by approximately 50 per cent (Pingali *et al.* 1997: 351). This infrastructure development to engineer the river, backed with the introduction of the market-oriented *doi-moi* policy in 1986, contributed to the delta being commonly described as the 'rice basket' of the country. Nowadays, 53 per cent of Vietnamese rice is produced in the Mekong delta, and this region has become a major focal point for food production (Sebesvari *et al.* 2012: 333). Moreover, while Vietnam had been a net importer of rice until the late 1980s, these initiatives have propelled the country into one of the world's top rice exporters.

Evers and Benedikter (2009) identified that the construction of new hydraulic infrastructure was, and still is, sustained by a close relationship among a number of key actors: central government ministries; provincial bureaucracies involved in irrigation and agriculture; and state-owned and private companies specializing in hydraulic construction and maintenance. Their study provides useful insight into how the Vietnamese hydrocracy comprised a range of stakeholders with influential decision-making powers that shaped the use of the Mekong waters at the national level. Such a constellation of actors indicates the power they have over the way the delta is engineered to maximize its potential for generating economic growth.

In explaining strategies and choices taken by these actors, Evers and Benedikter (2009: 417) maintained that hydraulic management centred on the control or 'taming' of the delta, rather than on averting water scarcity per se. However, I argue that real and potential water scarcity was none the less an important narrative for government bureaucracies in Vietnam, especially in framing water use in *multilateral* water institutions. The increased agricultural use of water led to concern about shortage during the dry season. Despite the many kilometres of canals that were constructed, storage capacity remained limited and there was infrastructure for only 2.8 per cent of the water from international transboundary rivers (Tu 2002: 342). Water shortage exasperates saline intrusion in the delta, as there is less freshwater flow to counter the salt water from the coast. Thus, water from upstream was needed, not only to maintain the irrigation systems but also to ensure that freshwater quantity could be maintained without it becoming saline. These concerns are repeatedly raised in speech acts in fora convened through the MC, IMC and MRC. Developing a discourse of water use through narratives of water scarcity was thus convenient in communicating downstream concerns.

An example is the Vietnamese reaction to the Kong-Chi-Mun (KCM) project proposed by the Thai hydrocracy in the late 1980s. The intension was to transfer up to 260 m³/sec of water from the Mekong River into the Chi-Mun basin in northeast Thailand for irrigation expansion (Tingsanchali and Singh 1996: 28). Vietnamese government officials publicly opposed the KCM project, claiming that water flow of 2,000 to 3,000 m³/sec was required to prevent saline intrusion (Hiebert 1991). There was also concern about another diversion project using two Mekong tributaries, the Kok-Ing-Nan project in northern Thailand. The Vietnamese Vice Minister of Water Resources argued that Thailand should consult other basin states prior to the commencement of feasibility studies (BBC 1994). For the Thai hydrocracy, these water transfer projects were domestic solutions to domestic water shortages in Thailand. For the Vietnamese hydrocracy, these projects could impinge upon water scarcity further downstream, and so 'rules' set up through multilateral water institutions would provide a way to monitor water use.

For the Thai hydrocracy, the narrative of water scarcity was no less significant and clearly drove the hydraulic mission. Historically, the domestic Chao Phraya River basin has been the key area for irrigation of the country. This basin has seen much hydraulic intervention through large-scale irrigation schemes and pumping stations, not to mention the construction of the Bhumipol and Sirikit dams, which provide major water storage. Dry season cropping has been advanced through such technological solutions. However, with its geographic location close to the metropolitan area of Bangkok, water demand is not exclusive to the agricultural sector. The Chao Phraya River basin is 'closed', with little dry season water availability to meet the multiple sector demands (Molle 2002; see also Molle 2007). The Kok-Ing-Nan project was devised as a way to increase water availability in this Chao Phraya basin by connecting the Nan tributary with the Kok and Ing rivers.

Hydraulic mission in the northeast region, where the KCM project was planned, has also led to the closing of river basins (Molle 2002: 2). Investment in various irrigation schemes and pumped storage underlines the need to avert water scarcity and, importantly, address issues of underdevelopment. The narrative of water scarcity in the northeast also chimed with political motives.

The aforementioned geopolitical intentions of the USA were targeted to this region to buffer against communist influences, thus resulting in both the central governments of Thailand and the USA allocating funds for development schemes. These strategic intentions were combined with the grand design of lower Mekong engineering, expressed through reports such as the Indicative Basin Report and in MC activities (Mirumachi 2012: 87–89). The KCM project was originally a project that would take advantage of water storage of the Pa Mong dam. However, once it became clear that the Pa Mong dam would not be implemented in the foreseeable future, the idea of an independent water transfer became the pragmatic solution to using the wealth of the river.

The DEDP played a crucial role in ensuring that the KCM project would be considered as part of the Mekong development initiative of the IMC, even if the Pa Mong may not materialize (Sneddon 2002). DEDP was one of the key agencies within the Thai hydrocracy and also represented the state in multilateral negotiations. This move exemplifies the bargaining power of the Thai hydrocracy. At the same time, the KCM project was to be complemented with internal plans for water supply in the northeast that also included smaller schemes of pumped irrigation (Mirumachi 2012: 90–91). Applying the FHH, it may be said that the Thai hydrocracy mobilized its financial and technical resources to increase their exploitation potential (Zeitoun and Warner 2006; see also Chapter 3, this volume). Diversified options for water supply within an advanced hydraulic mission, compared to that of Vietnam, strengthened its hydraulic control. The water scarcity narrative by the Thai hydrocracy is made material through these resources available to it, and this capacity to access and control water resources is comparatively stronger than that of Vietnam. Even though the KCM project has not been realized in its entirety as an international project, it may be argued that the Thai hydrocracy has been successful in buffering attempts to make water use rules disadvantageous for potential water transfer, as may be seen through the instance of securitization (see Sequence 5 of TWINS analysis). Through DEDP, the KCM project becomes the focal point of Thai state power that operates at multiple spatial scales. Not only is the project portrayed as an international agenda item, but also as a national plan (Sneddon 2002). However, domestically, the KCM project has been met with criticism and opposition by local communities, civil society groups and academics. These actors frame the KCM project as a local problem where dams constructed as part of the project have led to the inundation of land and forest resources, and subsequently loss of livelihood. There was criticism of the project profiting the political elite and not the local communities. Conflict over compensation between the state and local communities exemplifies a very different dimension and effect of the KCM project (ibid.: 2241–2244).

In summary, the domestic priorities for water resources use provide the context for the reasons why Thailand, as an upstream state, and Vietnam, as a downstream state, found water use rules contentious. Both states were going through the hydraulic mission where large-scale engineering projects were considered 'rational' by the hydrocracy. The demands of the *domestic* hydraulic mission manifested in water scarcity narratives, furthering the geographic imagination of the *regional* Mekong as an unharnessed yet promising river basin. Cooperation through the Mekong spirit is underpinned with notions of scarcity that justify various efforts of the hydraulic mission transposed at the international scale.

Water for energy: Xayaburi dam and the MRC procedures

The scalar politics of the hydraulic mission expressed through diplomatic notions of the Mekong spirit provide some interesting insight into one of the most debated topics in the basin currently: hydropower development. This topic helps illustrate how the historical development of water resources use and allocation has implications for water management practices now. As the following three sections will show, the institutional challenges of mediating interests over dam development necessarily include transboundary water interaction not just between Thailand and Vietnam, but also involve Laos making progress on mainstream dam plans, and a host of private sector actors. Moreover, the hydropower issue presents another dimension to how hydrohegemony is exercised.

One of the most notable developments in the Mekong River basin is the Xayaburi dam, currently being realized. Located in northwest Laos, the Xayaburi dam is part of the original Mekong Cascade proposed in the 1960s. After many decades of investigation, data collection and the accumulation of reports on the Mekong Cascade, this is to become the first-ever mainstream dam in the lower basin. This project has garnered regional and global attention, as it provides some insights into the potential future of the basin where there are a total of 12 mainstream dams being planned in the lower basin and 71 tributary dams to be fully operational by 2030 (ICEM 2010). In addition, by 2060 there could be over 130 dams (Räsänen *et al.* 2012: 3496). These dams represent the scale and geographies of energy demand, and the Xayaburi dam is intended to supply increasingly needed energy. This dam is designed to have an installed capacity of 1285MW. The Government of Laos signed a Memorandum of Understanding (MOU) with Ch. Karnchang Public Company, a Thai developer, in May 2007, and a project agreement a year later. In addition, the government signed a MOU for a Power Purchase Agreement with the Electricity Generating Authority of Thailand (EGAT) in 2010. These arrangements laid the ground plan for the transmission of approximately 95 per cent of the generated electricity to Thailand (International Rivers 2011). The Lao government initiated the first-ever PNPCA guided by the Mekong Agreement in September 2010. This process was contentious not only because of the opposition to the project, but also owing to lack of consensus around the process itself. The case of the Xayaburi dam exemplifies the interface between water and energy, and poses challenges to transboundary water governance through a multilateral RBO.

One of the most contested aspects about the Xayaburi dam was the extent of hydrological and ecological change which mainstream dams can cause. The reports of the environmental impact assessment and social impact assessment were eventually made available in the public domain, even though reports of this kind are often hard to obtain and not disclosed in the region. This reflects the degree of interest and scrutiny this project has received as the first main-stream dam in the lower basin. However, there was criticism about the rigour of the study, which only examined the impact of the project for 10km down-stream of the dam (International Rivers 2011). Consideration of fish biodi-versity was limited, at best, with sketchy analysis (Baran *et al.* 2011). The Vietnamese and Cambodian governments voiced grave concerns on the grounds of major impact downstream. In the formal response to the PNPCA initiated, the Vietnamese government expressed 'deep and serious concerns' that a range of negative impacts would be felt on the Mekong delta and that measures suggested by the Lao government would be inadequate (Socialist Republic of Viet Nam 2011: 2). The impacts particularly listed as potential concerns – 'saline intrusion, reduced fisheries and agricultural productivities, and degradation of bio-diversity' (ibid.) – are long-held concerns for down-stream Vietnam, not unique to just this project but to upstream water use in general. The Cambodian government raised similar socio-economic and environmental concerns downstream: 'fisheries, flow change, sediment balance, erosion, eco-system and agriculture land, livelihood' (Kingdom of Cambodia 2011: 2). The Thai government argued that 'the sustainability of the project is still questionable', with concerns of biodiversity loss and impacts upon local livelihoods, and sedimentation. The level of impact assessment and proposed mitigation measures were also found wanting (Kingdom of Thai-land 2011: 2).

To aid the PNPCA process, the MRC commissioned a Strategic Environ-mental Assessment (SEA) of planned mainstream dams in the lower Mekong. The findings of this study concluded that there would be irreversible ecolo-gical impacts and damage to fisheries, which would lead to loss of livelihood and food insecurity. The study advised that mainstream projects be postponed for ten years, while alternative options could be considered and decision-making capacity developed (ICEM 2010). In line with the recommendation made by the SEA report, the Vietnamese government called a halt on main-stream dam development for ten years. As a way of exercising bargaining power, the Cambodian government attempted to pressurize the Lao govern-ment as a downstream bloc with Vietnam (Worrell 2012a, 2012b). In con-trast, the Lao government argued that the Xayaburi dam is a necessary solution to the increasing energy demands of the region, benefiting not just the country but also the Mekong region as a whole. Moreover, the concerns of impacts upon biodiversity could be addressed with technological measures such as improved fish passages. The Xayaburi dam is stressed as a run-of-river dam, thus having minimal impacts upon sediment and river flow (Lao PDR 2011).

The PNPCA provided some scope for discussions about the design of the dam and downstream impact minimization measures, such as fish passages. However, this process was controversial in the sense that it did not result in the uniform action of succeeding steps. MRC member states failed to reach a consensus and referred the matter to the Ministerial Council, the highest decision-making forum within the MRC, in April 2011. By the end of that year, the Ministerial Council had provided no firm decision on the dam and suggested further studies be carried out in order to identify the impact of mainstream dams. None the less, the Lao government had already declared the PNPCA process to be completed in June 2011. This put into motion the signing of the Power Purchase Agreement with EGAT and green-lighted construction. In November 2012 the project officially began, despite some claims that the Lao government had violated the Mekong Agreement (see Herbertson 2013).

The development of mainstream dams posed many unknowns to decision-making within the MRC. There are high levels of uncertainty over the extent and ways in which impacts to ecosystems will be addressed, not to mention the socio-economic impacts. Importantly, there was no consensus reached through the RBO on the best way to address these impacts while achieving economic development by using the river. This uncertainty is portrayed in the multilateral process of applying the PNPCA under the Mekong Agreement for the first time. Because construction went ahead without formal PNPCA agreement, it may be said that the political process did not provide a strong mechanism in which different water use interests and river basin concerns could be mediated. The differences between Laos and the downstream states exemplify the malleable interpretation of sustainable development under the Mekong Agreement. While there is no acute military conflict over the Mekong waters, politicization of shared waters coexists with institutional cooperation.

Power and profits

The Thai government did express its reservations during the PNPCA process, as explained above. However, a closer look at the actors involved in hydro-power development begins to reveal vested Thai interests. The aim of this section is to show how this coexistence of conflict and cooperation is made up of a complex political economy of water and energy that accentuates hydro-hegemonic control over water resources as well as the economic returns of exploitation. Moreover, this set-up did not materialize overnight, but through several decades of policy reform and new actors being associated with the hydrocracy.

The export of hydroelectricity from Laos to Thailand already had a precedent: the Nam Ngum Dam Project. This Mekong tributary project on the

Nam Ngum River was completed in 1971, as one of the cooperative, regional development initiatives under the MC. While it only had an initial hydropower capacity of 30 MW – much smaller than dams envisaged for the Mekong Cascade – this project was symbolic for the MC. Since then, the Theun–Hinboun, Houay Ho and Nam Theun 2 hydropower projects have been developed, as well as the Nam Ngum 2 dam, augmenting the hydropower capacity of the Mekong tributary rivers. These projects have been facilitated by several MOUs between the two governments on the generation and sale of energy that has steadily increased the capacity for export.

All hydropower from Laos is sold to EGAT, being the only authority for energy generation and transmission in Thailand. Moreover, EGAT has shares in Independent Power Producers (IPPs), private companies for energy generation that have invested in hydropower projects in Laos. The unique position that EGAT holds arises out of liberalization efforts of the energy sector that has been derailed and deformed over time. Triggered by loan conditionalities from the World Bank in the early 1980s, the government embarked upon a plan to liberalize the energy sector and privatize state-owned enterprises. The introduction of IPPs and Small Power Producers (SPPs) in 1992 was a way to allow privatization in energy generation. The interest in investment by IPPs was high. However, the IPPs established during this early phase of privatization, the Electricity Generating Company Limited (EGCO) and Ratchaburi Electricity Generating Holding Public Co Ltd (Ratchaburi), were criticized for posing a conflict of interests. For example, EGCO was set up as a subsidiary to EGAT but with EGAT having approximately 40 per cent shares in the company at that time and being the only organization that can buy generated electricity, it could pay for higher energy prices to boost company profits at the expense of consumers (Greacen and Greacen 2004: 523–524; see also Matthews 2012: 398). Although there were plans to corporatize EGAT from a state-owned enterprise as a major feature of the energy sector reform, this was ultimately rejected by the Thaksin government which came to power in 2001. Instead, the Enhanced Single Buyer scheme was introduced, allowing EGAT to continue to be the sole authority for generation and transmission. EGAT, with a current shareholding of 25.41 per cent and 45 per cent in EGCO and Ratchaburi respectively, has been criticized as being a monopoly sustained by inadequate laws and regulatory mechanisms for a competitive market (Wisuttisak 2012).

Historically, Thai energy actors have had vested interests in water resources development in the Mekong. EGAT, a state-owned enterprise in the energy sector, has implemented national dam projects, such as the aforementioned Bhumipol and Sirikit dams in the Chao Phraya River, since they have a hydropower component in addition to providing water storage. The multilateral negotiations for river basin development have been led by the National Energy Authority and DEDP during the MC and IMC periods. EGAT's interest in Mekong waters has grown, as domestic opportunities to develop hydropower projects are now virtually non-existent. Strong civil society opposition to dam development has made the option of new development difficult, and instead opportunities abroad, such a Laos, Myanmar and the Yunnan province in China, have been considered (Greacen and Palettu 2007: 97–98). Hydropower plans in places like Laos and other parts of the Mekong are facilitated by investment schemes of build–own–transfer and public–private partnerships. These schemes tend to play up profits for the investors and the host country, and play down risks and impacts (Middleton *et al.* 2014). Past dam development projects have shown that the network of financial flows between the private sector actors, power companies, host government and financial institutions are incredibly complex, and deliberately so. Merme *et al.* (2014: 26) pointed out that complex financing in itself is becoming a key service that motivates private sector actors to engage in dam development, rather than energy prospects themselves. The financial structure of large-scale projects, and their lack of clarity and disclosure, raise serious questions about accountability of actors (ibid.).

It should also be mentioned that the Lao government has been keen to develop hydropower. Since the early 1990s, the Lao government prioritized hydropower development for energy export as one of the ways in which to attract foreign investment after introducing a market-oriented economy structure in 1986. In the eyes of the Lao hydrocracy, hydropower export is an efficient way of using water resources for economic growth. The Xayaburi dam project is a clear example of the Lao government using its natural resources to generate economic development. The transboundary relationship among the EGAT, private developing and financing companies, and the Lao hydrocracy creates a political economic structure where hydropower development is considered desirable (Matthews 2012).

Moreover, energy development is being spurred on by regional integration ideals through energy grids as part of the Greater Mekong Subregion Programme by the ADB, and is also discussed within the Association of Southeast Asian Nations (ASEAN). In addition to private energy companies, regional banks and construction companies are 'new' actors that are driving hydropower development, sidelining 'old' actors such as development banks, IFIs, UN agencies and bilateral donors (Middleton *et al.* 2009). The elite decision-makers of the hydrocracy who have direct channels to multilateral discussions on issues like the Xayaburi dam are part of a group of actors that straddle both private and public sectors concerned with water and energy.

Hydro-hegemony through energy development

Just as water scarcity may be socially constructed (Mehta 2003; see also Chapter 1, this volume), so too may energy need. Analysts point out that energy demand projections by the Thai government have been overestimated over the years in an attempt to justify government expenditure and investment in energy projects (Greacen and Greacen 2004; Graecen and

Palettu 2007). Hydropower contributes 5 per cent to the energy mix, much less compared to natural gas and coal (EGAT 2011). However, with its seemingly 'green' credentials, hydropower helps diversify the energy mix. Electricity from Laos, as well as from Malaysia, amounts to 2,184.60MW, contributing just under 7 per cent of energy supply in Thailand in 2011 (ibid.). The inflated energy projections and modest contribution of hydropower from abroad present some interesting justifications of investing in Lao hydropower by the hydrocracy and elite decision-makers. A common reason was suggested as helping a poorer neighbour and contributing to regional development (Thai governmental official 2013, pers. comm.; Thai energy professional 2013, pers. comm.; Thai energy analyst 2013, pers. comm.). In contrast, analysts critique that the socio-political climate of countries like Laos prioritize economic development over ecological and socio-economic concerns, limiting dissenting voices. This makes project development seem more convenient than at home for Thai investors and elite decision-makers (Matthews 2012; Simpson 2007). In addition, Thai construction companies are already heavily vested in the hydraulic mission abroad and it has been pointed out that this structure in particular provides returns for the Thai economic elite (Middleton *et al.* 2009; Jensen and Lange 2013: 50).

This contradiction exposes the hydro-hegemonic control Thailand has over water resources in the Mekong. The rhetoric of regional cooperation may seem to be a use of 'soft' power that integrates the interests of both the hydro-hegemon and non-hydro-hegemon (Zeitoun *et al.* 2012). However, in effect, it commodifies 'external' water resources to its territory as something that may be bartered with the hydro-hegemon's advantageous material capability. It is not only the financial strength of the state and private investors that strengthens material capability, but also the structure of the energy sector that supports EGAT and IPPs. Equating water resources as exploitable goods conjures a geographic imagination of Laos with untapped wealth that provides a win–win situation across the borders. Barney (2009) pointed out that Laos may be framed as a resource frontier where neoliberal ideas of markets and private investments are welcomed within a communist regime. Added to this are opportunist hydro-hegemonic interests. The Thai hydrocracy views projects in Laos to be profitable *and* less threatening than the issue of water flow regulation through Chinese mainstream dams. The Manwan and Dachaoshan dams contribute to the so-called Yunnan Cascade of eight mainstream dams in the upper Mekong basin. There are also tributary projects in the pipeline which will significantly increase the total water storage capacity in the upper Mekong basin. There are views that the development of the Mekong River rests on upstream Chinese dam development and that China has real control over the water (Thai Governmental Official B 2008, pers. comm.; Former Thai Governmental Official 2008, pers. comm.).

These concerns suggest that China has hydro-hegemonic control over the *entire* river basin, based on its geographic location and ability to mobilize funds for hydraulic development. Even though the suggestion of including

China and Myanmar in the river basin organization was rejected in the process of establishing the MRC, the Thai hydrocracy continued to emphasize the importance of upstream states' involvement.[5] Interviews indicated that the Thai hydrocracy seemed to be more preoccupied with engaging China than applying principles of sustainable development within the MRC framework. The river basin institution gradually becomes less of a political forum to exercise its power on rule-making over water resources. For example, it was suggested that Thailand no longer relies on the MRC for its water resources development (Thai Governmental Official A 2008, pers. comm.). Framing Chinese dams as a threat takes the pressure off justifying its support and demand for hydropower development in Laos.

The political economy and the discursive framing regarding the importance of dams secure the hydro-hegemonic status of Thailand over benefit derived from water use in the *lower* basin. This is an important point which shows that hydro-hegemony is relative, in this case based on the unit of governance: basin-wide or part of it. However, in the domestic politics sphere, the rhetoric posed on hydropower development is increasingly challenged. There has been much criticism over the Xayaburi dam, and many Thai civil society groups mobilized campaigns to advocate the wide-ranging impacts it can cause. Protest letters targeted the Thai prime minister and local politicians (Thai Mekong Community Network of 8 provinces 2011). In addition, there have been legal attempts to halt the project via a court ruling over EGAT and other governmental organizations (WWF 2014). These civil society groups have formed coalitions with other regional and international NGOs to influence different fora. For example, Save the Mekong brings together key regional organizations and International Rivers, a prominent international NGO with experience of campaigning against dams globally. The network of these groups enables them to access different nodes of decision-making, not just at the local and national government levels, but also at the MRC and even ASEAN levels. When examining hydro-hegemony from a spatial perspective (Warner 2008), the Thai hydrocracy has faced various levels of opposition to its hydro-hegemonic control at different spatial scales.

Even though the construction of the Xayaburi dam has proceeded, the Vietnamese hydrocracy has continued to caution against mainstream dams. In a diplomatically worded speech, the Vice Chairman of the Viet Nam National Mekong Committee, Nguyen Thai Lai, asserted:

> each riparian country should show their responsibility by assuring that any future development and management of water resources proposed in the basin should be considered with due care and full precaution based on best scientific understanding of the potential impacts.
>
> (Lai 2013: n.p.)

However, this caution over mainstream dams also reveals a contradiction in the Vietnamese perspective on dam impacts. In Vietnam, the Yalli Falls dam is one of the controversial dams on the Sesan River, a tributary of Mekong River, shared with Cambodia. Many ecological and socio-economic concerns have been raised, as the 760 MW hydropower dam changed the flooding pattern in the area and fish species have declined. The impact upon livelihoods for local communities has been experienced in both Vietnam and downstream Cambodia. These impacts are largely attributed to the lack of robust assessment and mitigation measures. The EIA has been criticized as being narrow in its scope of assessment, ignoring the impact downstream in Cambodia (Wyatt and Baird 2007). As this project was constructed in 1995, before the Mekong Agreement, there were no legal obligations to notify other basin states about the project. However, this unilateral practice has set a precedent for other hydropower projects in the Sesan River (ibid.).

Despite expressing its opposition to the Xayaburi dam, the Cambodian government also has stakes in hydropower development in the basin. It has taken a pragmatic approach to its renewed membership of the MRC, as the organization offers access to much-needed technical and financial assistance, which facilitates water resources development projects. It has also accepted Chinese investment in large-scale infrastructure development, including hydropower in recent years. In an attempt to entice investment, the Cambodian government has been reticent in dealing with the environmental impact of Chinese dams (Keskinen *et al.* 2008). In addition, it is preoccupied with its own mainstream dams as electricity demand, both in the country and the region, is expected to rise. The Mekong Cascade still has the allure for this downstream hydrocracy, with plans to develop the Sambor Dam. A MOU was secured in 2009 between the government and a Vietnamese developer for the Stung Treng dam. Although overt conflict is unlikely to occur in the circumstances outlined above, high-intensity cooperation to fundamentally address the socio-economic and environmental impact of hydropower development remains difficult to achieve.

Who governs the Mekong?

A detailed analysis of the transboundary water interactions between Thailand and Vietnam has demonstrated the ways in which water is politicized, and even securitized, combined with varying cooperation intensities. The analysis reflects Menniken's (2008: 299) claim that '[b]ecause water contains conflict, but is susceptible to cooperation, it is no surprise that both events occur concomitantly as strategies of riparian actors in transboundary river basin'. However, by using the TWINS framework which focuses on transboundary water interactions rather than events of conflict and cooperation, my analysis further explained how and why conflict and cooperation coexisted. It was revealed that national interest, in the guise of 'cooperation', has been a recurring factor since the early days of the river basin organization. The convergence between the hydrocracies on narrowly defined national interests of macro-economic development played a significant role in sustaining the existence of the river basin institution. What the Mekong spirit reflects is in fact this skewed perception of water use and allocation. Under the surface of this Mekong spirit there is tension over the implementation of the norms of the river basin institution, borne out of different water use priorities. The scalar politics of domestic water demands and policy ideals translating into inter-state negotiations are mediated through the hydrocracy. These elite decision-makers are endowed with different material capability and discursive power. The Thai hydrocracy has been effective in using its material capacity to secure options for water storage and use. In addition, discursive power was used at various river basin fora, to consolidate its control over the shared waters in a way that best suited its water development priorities. Consequently, the MC, IMC and MRC have been useful vehicles to transpose interests of the hydrocracy and elite decision-makers on to a broader canvas of the Mekong River, drawn up with geographic imagination. Thai hydro-hegemonic control is further strengthened by another dimension of water resources control, through hydropower development.

The hydropower boom in the region begins to pose some very serious questions on who governs the Mekong. National interests merely prioritize economic development through the exploitation of the river, when in fact such development has serious impacts at the local scale (Hirsch *et al.* 2006). In the pursuit of national hydraulic missions, binding rules have been rejected. As the Xayaburi dam showed, interest to exploit Lao water resources is driven by a complex set of state decision-makers from water and energy sectors, private investors and construction companies connected transnationally. This political economic structure, as well as hydropower interests of Vietnam and Cambodia, highlights the implications of institution-building efforts based on the Mekong spirit. The 'ambiguous' nature of the Mekong spirit (Hirsch *et al.* 2006: 53) suppresses acute conflict but also limits cooperation, demonstrated by the lack of action to address the transboundary environmental impact of mainstream dams. The MRC has been faulted as being ineffective as a result of a limited mandate when it comes to dealing with hydropower issues in the past (Wyatt and Baird 2007: 438–439). Dore and Lazarus (2009) warn that the MRC could become redundant, having experienced so many shortcomings in facilitating wider deliberation across a range of stakeholders. Yong and Grundy-Warr (2012) critiqued that if alternative discourses by civil society – and a consequent re-imagining of the Mekong – are not taken into consideration, the issue of hydropower development could well be apoliticized. National interests continue to be the concern of only a small number of stakeholders, which have the political and economic power to exploit the water without having to experience the adverse impact first hand. The role of the hydrocracy and the clique of elite decision-makers have a large impact upon Mekong water governance. However, on the question of who *should* govern the Mekong, it has become increasingly clear that the elite decision-making is problematic and a wider range of stakeholders needs to be involved. The challenge will be to consolidate very different worldviews of stakeholders, not all equally empowered. However, a first step would be to address the account-ability of existing decision-making procedures. Drawing upon insights from this case study and also from the other two basins, the next chapter further considers the resilience of elite decision-making to alternative discourses and measures for accountability.

Notes

1 Later succeeded, in 1974, by the United Nations Economic and Social Commis-sion for Asia and the Pacific (ESCAP).
2 Keen to become influential in the Mekong region so as to suppress communist influences, the USA convinced the four states to request an American study, instead of one by ECAFE (Sangchai 1967: 37). However, the four states preferred the pres-ence and facilitation of ECAFE in the river development enterprise (Makim 1997: 75).
3 North Vietnam is not within the basin and was excluded from membership of the MC because it did not belong to ECAFE. It was also not recognized as a sovereign state by the other basin states (Makim 1997: 73).
4 Article 26 states:

> The Joint Committee shall prepare and propose for approval of the Council, inter alia, Rules for Water Utilization and Inter-Basin Diversions pursuant to Articles 5 and 6, including but not limited to: 1) establishing the time frame for the wet and dry seasons; 2) establishing the location of hydrological sta-tions, and determining and maintaining the flow level requirements at each station; 3) setting out criteria for determining surplus quantities of water during the dry season on the mainstream; 4) improving upon the mechanism to monitor intra-basin use; and 5) setting up a mechanism to monitor inter-basin diversions from the mainstream.
>
> (Agreement 1995: Article 26)

5 Since 1996, China and Myanmar have been dialogue partners which participate in MRC meetings without decision-making powers.

7 Making sense of transboundary water politics

Seeking determinants of conflict and cooperation

So what do the three case studies of the Ganges, Orange–Senqu and Mekong river basins tell us about coexisting conflict and cooperation? What makes water allocation and use political? At the beginning of this book, I set out to explore how, and why, conflict and cooperation occur during the process of addressing transboundary water issues. The conceptual development of TWINS was an attempt to address critiques of theoretical underdevelopment within the study of hydropolitics. The existing literature, often drawing upon IR and political science, relies on a binary of conflict and cooperation to characterize river basins, resulting in a reductionist interpretation of hydropolitics. Such an approach misses out on issues of soft power and on the consideration of spatial dimensions of water demands and policy-making. These serious deficits hinder understanding of wicked, messy, fluid and power-laden processes of international decision-making over shared rivers.

Providing a comprehensive explanation of conflict and cooperation required an inquiry not bound by disciplinary silos related to conceptualizations of conflict and cooperation. It necessitated thinking extending beyond the ideas developed through IR and political science. I used the concept of transboundary water interactions to help explain the politics that surround shared waters. Examining transboundary water interactions allows for a better appreciation of the changing, evolving nature of relationships among basin states. Analysis using the TWINS matrix not only indicates the existence of both conflict and cooperation, but also the specific degrees of coexisting conflict and cooperation. The analysis I have presented in the previous chapters thus breaks the mould of classic hydropolitics which examines moments or events of conflict and cooperation. This is the reason why I call this a study on transboundary water politics, in an attempt to emphasize the discursive process through which water becomes political, rather than on conflict or cooperation as outcomes.

As an original approach, the TWINS framework enabled a fresh look at when and why coexisting conflict and cooperation intensities change between basin states. The focus on power showed that hydro-hegemonic control over water resources is exercised not through overt means of military force, but through the ways in which agreements are tabled and projects green-lighted or delayed. The interdisciplinary nature of the TWINS framework also allows explanations to be found in geopolitical developments, geographic imagination of the river basin and policy interfaces with other sectors such as energy. These factors are often ignored in studies that look at either conflict or cooperation over waters.

The TWINS framework is useful to pinpoint the discursive power of specific agents. In particular, I have focused on elite decision-makers made up of the hydrocracy and politicians supporting the hydraulic mission of this institution, and their speech acts that socially construct transboundary water interactions. The aim was not to privilege the hydrocracy and elite decision-makers as prime and sole actors that construct transboundary water politics. Rather, the intention was to shed some light on what often gets generalized as the 'black box' of decision-making, and to provide insights into their interests and means of influence, both at the domestic and international levels. Agencies of elite decision-makers and their power are manifested in a political economy that shapes the demands of water resources and types of management mechanisms. 'National' interests are played out in international fora; scales of managing the river are constructed to suit these interests. The TWINS framework incorporates a critical perspective of scalar politics played out by these elite decision-makers.

In this chapter, insights from the empirical studies are brought together to further refine some of the ideas on power, scale and agency. After summarizing key characteristics of the transboundary water interactions of the three case studies, I first draw connections between geographic imagination and material capability. By doing so I review and refine the FHH to better explain how power is exercised. Second, a critical account of 'shallow cooperation' is provided. The case studies using the TWINS framework provided an in-depth analysis of why cooperation in transboundary river basins has been limited and patchy (e.g. Conca *et al.* 2006; Dombrowsky 2007). Understanding how hydro-hegemony works in contexts of shallow cooperation provides important insights into issues of equity of water use and allocation. Third, I shed light on the nature of elite decision-making that imposes a very particular problem frame to deal with transboundary water management. The case studies revealed major problems and contradictions of choices borne out of the hydraulic mission. The effect of a dominant problem frame poses significant challenges, particularly for multilateral water governance which is repeatedly promoted through policy initiatives. Finally, the challenge of practising transboundary water governance is explored.

TWINS and degrees of coexisting conflict and cooperation

The geographer, Gilbert White, commented almost half a century ago that '[i]f there is any conclusion that springs from a comparative study of river systems, it is that no two are the same' (White 1957: 160). The Ganges, Orange–Senqu and Mekong river basins all tell a different story of how transboundary water interactions have developed, and the implications for transboundary water governance. The results of the trajectories on the TWINS matrix showed different dynamics of upstream–midstream–downstream interactions in these three basins. The bilateral relationship between Nepal and India was the most static, while the relationships between Lesotho and South Africa, and between Thailand and Vietnam, were prone to more frequent changes of conflict and cooperation intensities. The South Asian case study showed fewer speech acts that dramatically changed conflict and cooperation intensities, for two reasons. To emphasize: this does not mean that the Ganges case study was a better or good basin. As was pointed out in Chapter 4, this static nature does not necessarily indicate sustainable or equitable water allocation and management, or that there were better governance mechanisms in place. In this case, static transboundary water interactions may be attributed to, first, lengthy negotiations without any immediate results, and second, political stalemate over decisions related to project implementation after a major treaty was signed in 1996. It is interesting to note that none of the river basins experienced situations of high conflict and high cooperation, an unstable and unsustainable form of relationship, as explained by Craig (1993).

The findings of coexisting conflict and cooperation substantiate existing empirical studies of the Ganges, Orange–Senqu and Mekong river basins that identified signs of coexisting conflict and cooperation. For example, Elhance (1999: 187) summarized that the hydropower potential of the Himalayan region provided opportunities for cooperation between Nepal and India, but in fact brought about a contentious political relationship, hindering water resources development. By examining water agreements between states in Southern Africa, Kistin (2007) argued that the implementation of agreements provides the opportunity for both conflict and cooperation. Similarly, Öjendal (2000: 283) noted that in the lower Mekong region, '[t]he pattern of conflict and cooperation that crystallizes is thus one of higher conflict potential, in combination with an ambition to play down the level of conflict'. That water resources may be the subject of both conflict and cooperation between local communities has also been demonstrated in studies such as Wade (1988), Wessels (2007), and Komakech and van der Zaag (2011), drawing upon contexts from India, Syria and Tanzania respectively. My analysis of the three case studies enabled by the TWINS framework provides further qualitative evidence that the same may be said for international transboundary river basins. Conflict does not necessarily exclude cooperation, and vice versa, when it comes to allocating and using transboundary waters and deciding on ways to manage the river.

The TWINS matrix proved its usefulness to illustrate and communicate the change of coexisting conflict and cooperation that cannot be captured in one single intensity scale. The TWINS matrix is a new analytical tool to probe the nuances of transboundary water interactions that forces the analyst to constantly consider the 'mixture of conflicting and complementary interests', pointed out by Axelrod and Keohane (1985: 226). In addition, it provides an example of how speech act theory may be applied. Insights from these transboundary water interactions begins to explain the challenges of basin-wide governance. Making this link is important, especially as policy promotes basin-wide and regional attempts at water governance (DFID and WWF 2010; Royal Academy of Engineering 2010; Jensen and Lange 2013; UN Water 2013b). To this end, the TWINS matrix may be applied to a range of relationships not limited to state actors. For example, the matrix may be used at the subnational scale, between the government and civil society, or to trace the evolution of relationships of transnational advocacy networks, as conceptualized by Keck and Sikkink (1998). Moreover, the TWINS matrix may be used to examine multiple sets of bilateral relationships that may then be synthesized to examine one basin, or to plot multilateral relationships. The current use of the TWINS matrix serves the purpose of furthering our understanding of an opaque decision-making process by elite decision-makers regarding water use and allocation. In this study, one pair of states was initially selected in each basin to examine transboundary water interactions using the TWINS matrix. It is acknowledged that these bilateral transboundary water interactions do not occur in a vacuum, in isolation from other basin states. Focusing on two basin states was a way to demonstrate in more detail how interests play out in the form of hydro-hegemony and its implications for basin-wide governance.

Transboundary water interactions defying the divide of low and high politics

As a first step towards explaining why intensities of coexisting conflict and cooperation changed, the TWINS analysis showed that transboundary water interactions are bound up in broader political developments. Importantly, the pace of negotiations is influenced by geopolitical and diplomatic contexts. In the Lesotho–South Africa relationship, negotiations and feasibility studies of the LHWP were occasionally halted, as a result of increasingly tense bilateral relations during the 1970s and 1980s. While common goals of development for the water transfer scheme were established, specifying rules and procedures of the project was difficult to achieve. In the Mekong River basin, tension over political ideology within the region led to the quadrilateral basin institution, the MC, to be redundant without representatives between 1976 and 1977. The impact of geopolitics is also evident in the Ganges River basin. Although it cannot be claimed that deteriorating political relations was the only factor which delayed negotiations, in the Nepal–India case geopolitical and security issues contributed to a political climate that was not conducive to expedited agreements over hydraulic development from the 1960s to the 1980s.

Martin *et al.* (2011) pointed out that transboundary environmental management as low politics forms organically with its interaction with high politics. Their study, using an adapted version of the TWINS matrix for transboundary ecosystem management, showed that states do cooperate despite regional violent conflict. Low politics of biodiversity conservation between the Democratic Republic of Congo, Rwanda and Uganda is not excluded from deliberation in situations of contested high politics. They argued that low and high politics influence each other, and that 'the nexus of their trajectories is not based on simple, linear causality but on a more complex constitution of management spaces, featuring a dialectic between institutions and relationships' (ibid.: 632). Drawing their findings on to contexts of transboundary river basin management, the case studies in this book show how water issues also become a geopolitical agenda item. In other words, water issues, traditionally considered as a low politics issue, can transform into being part of high politics. Thus this 'nexus' is one where low and high politics can become difficult to differentiate. Following this, I argue that categorizing transboundary water interaction as one limited to low politics, which may occasionally intersect with high politics, falls short of fully explaining the politics over shared waters.

An extreme example of water use transforming into high politics is found in instances of water securitization. In the case of securitization of the LHWP explored in Chapter 5 (this volume), this water transfer project embodied South Africa's geopolitical and regional security concern. Elite decision-makers viewed the project as benefiting the hydraulic mission and as a way for some kind of working relationship with Lesotho in an increasingly politically instable region. The project development was at once subject to issues of national security *and* a medium through which national security could be enhanced in tense geopolitical settings. Gleick (1993: 87) argued that water resources can have strategic military value in times of (geo)political conflict, especially as they are effectively 'finite, poorly distributed and often subject of substantial control by one nation or group'. However, it is not so much the scarcity of water resources that lends itself to becoming targets or tools of military intervention. Rather, it is the strategic political and economic interests arising from exploiting the water resources that matter. In the Mekong case study, the restriction of water use by the RBO impinged upon national development plans for Thailand, perceived equal to interference in sovereignty by decision-makers. These grave reservations led to securitization, as examined in Chapter 6 (this volume). Rather than a hierarchical relationship between high and low politics, securitization of water shows that it may be prioritized and de-prioritized within national agendas. Explaining the evolution of transboundary water interactions over time requires better examination of the tools and means through which water issues are framed and negotiated. Binaries, in this case of low and high politics, fail yet again to explain transboundary water politics (see also Chapter 2, this volume). In the next section we turn to the ways in which material capability lends to the mixing of low and high politics.

Hydro-hegemony and material capacity

By virtue of their comparatively stronger economic capability, the hydro-hegemonic states of South Africa, India and Thailand were able to initiate water-capturing activities within the basins. Notably, comparatively stronger material capabilities made it easier for the hydro-hegemons to implement or to follow up securitizing/opportunitizing moves, which meant they could achieve the kind of hydraulic control that best suited their water resource demands. Based on the examined cases, I argue that the level of material capability, which defines the pace of hydraulic and institutional development, plays an important role in the mix of low and high politics in certain instances of transboundary water interactions.

To elaborate, the South African government had the economic capability to mobilize financing for the construction costs of the LHWP in Basotho territory. The project was instigated by the South African hydraulic mission, which framed water transfer projects as part of the solution to addressing water scarcity. Securitizing speech acts concerning the LHWP were credible because the government had comparatively stronger military power to secure the border with Lesotho in 1986. Similarly, the Indian government's comparatively stronger economic capability allowed it to fund and construct hydraulic projects in Nepali territory, such as the Kosi and Gandak projects. In addition, it proposed to Nepal that large-scale projects − such as the Pancheshwar and Kosi High Dam projects − should be constructed, in an attempt to ensure irrigation, hydropower and flood control benefits. Strong comparative material capabilities facilitated India's unilateral action to construct part of the Tanakpur project in Nepal. This is analysed as an opportunitizing speech act in Chapter 4 (this volume). In the case of Thailand, the economic capacity and engineering resources to investigate and design unilateral water diversion projects were instrumental after it became clear that the multilateral water management institutions within the Lower Mekong River basin would not facilitate mainstream projects as quickly as originally planned. In addition, the securitizing move in 1992 was underpinned by material capability that made the idea of an alternative forum involving all six states for a new water management institution (that would safeguard its water utilization plans) a not completely unrealistic one.

Importantly, material capability to mobilize funds and human resources, and to conduct scientific studies, has enabled alternative options for water resources development to be considered when bilateral/multilateral arrangements did not suit the interests of the elite decision-makers. For example, the South African hydrocracy hesitated to implement a second phase of the LHWP in the late 1990s because water resources demand was not as high as expected. Rather than committing to a project that could be a potentially unnecessary investment, the South African hydrocracy investigated a domestic water transfer project, the Thukela Water Project, as an alternative, so stalling bilateral negotiations. In the Mekong case, by the time a new cooperative framework was negotiated in the early 1990s among the four lower Mekong basin states, the Thai hydrocracy was less reliant on the multilateral water management institution for investigating and implementing hydraulic projects. It had enough financial and human resources of its own to plan unilateral hydraulic development options in its northeast regions to abstract water from the Mekong River system. With regard to India, the hydrocracy investigated the River Linking Project, designed to divert water from the Himalayan region. It was potentially one way in which the Indian hydrocracy could increase water supply without having to engage in bilateral or multilateral discussions. By having an alternative option, India could postpone decisions on the implementation of the Mahakali Treaty while considering its own hydraulic mission.

The FHH suggests that the riparian position and the potential to exploit water resources, along with power, enable advantageous control of water resources (Zeitoun and Warner 2006). As was explained in Chapter 3 (this volume), the riparian position and exploitation potential represent the material capability of basin states to seek access and control of water resources. In other words, using comparatively stronger material capabilities or discursive power, or a combination of the two, enables hydro-hegemony. Thus, hydro-hegemony is not simply about immaterial power. This material capability has the important effect of materializing geographic imagination. For example, the Mekong River basin was seen as an underdeveloped region with untapped potential for economic growth. To the South African hydrocracy, the Orange–Senqu River was an untamed river that needed to be managed for the benefit of society. The LHWP was a 'jewel' that symbolized the efforts of regulating nature for improving economic standards and a rebirth of a region, as expressed by the South African President, Thabo Mbeki (Mbeki 2004: n.p). Hydropower projects proposed in the upper Ganges River basin between Nepal and India draw up a landscape for dam engineering. It has been argued that making geographic imagination real is 'contingent on technology and scientific expertise, human labor, capital investment, and social and political power' (Kaika 2006: 277). The case studies in this book affirm that engineering expertise and capital are certainly important. At the same time, it shows that when producing such imaginations at the transboundary level, material capability enables and supplements the effective use of soft power, making it an important consideration in the analysis of transboundary water interactions.

From the perspective of elite decision-makers with vested interests in the hydraulic mission, a wider set of hydraulic development options enables states to become water secure. The hydro-hegemon's prerogative in choosing when to engage in cooperative initiatives is explained by Zeitoun *et al.* (2011) as relating to soft power, or the discursive means of exerting influence over decisions. However, the extent to which basin states can capitalize on their riparian position, engineering skills, scientific expertise and financial resources (including attracting investors and donors) is not insignificant. They provide leverage to the claims made on the need for water resources and threats to and from the river basin. These needs and threats may be socially constructed and not necessarily real, but the material capability enables the discursive framing of the hydro-hegemon more credible. Soft power in hydro-hegemonic settings is all the more effective with material capability.

Shallow transboundary water interaction with an ugly face

The analysis, enabled by the TWINS approach, sheds light on what 'shallow cooperation' over water resources entails. The 2006 UNDP Human Development Report, entitled *Beyond Scarcity: Power, Poverty and the Global Water Crisis*, stated that there is recurring shallow cooperation and little deep cooperation (UNDP 2006). Shallow cooperation occurs when the scope of cooperation is limited and there are many issues, reflected in institutional arrangements and provisions (Gerlak and Grant 2009). This leads to a situation where cooperation fails to meet wider human development goals, brushing aside issues of environmental degradation and impacts upon livelihoods (UNDP 2006: 222). Studies on the global status of international transboundary river basins, such as those by Wolf *et al.* (2003), show that rivers are being managed within shallow cooperative modes rather than conflictual modes, and that some basins have institutionalized mechanisms in place for water resources management, though they vary in scale and scope. Shallow cooperation, re-conceptualized through the TWINS matrix, would be applied to situations where, for example, basin states are engaged in conflictual interactions of medium intensity *and* cooperative interactions of low to medium-high intensities. In other words, common norm may be formed on water allocation through a treaty, but implementation is, none the less, politicized (see Figure 7.1).

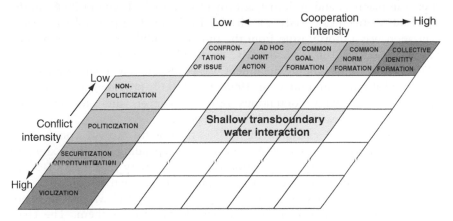

Figure 7.1 Shallow transboundary water interaction within the TWINS matrix.

Understanding shallow transboundary water interaction uncovers the way in which elite decision-makers project their interests on negotiations over water. In the 2006 UNDP Human Development Report mentioned above, it was suggested that asymmetrical negotiating power can act as a disincentive to collective action. In addition, sovereignty confounds how rivers are to be shared, bringing about situations where national agendas for water resources development are absolute and regional plans dismissed (UNDP 2006: 223). The TWINS analysis indicated that power asymmetry matters, and that sovereignty and national agendas do get in the way. However, the effects of power asymmetry may be qualified further than the generalized claim about inhibiting cooperation. Power asymmetry does not prohibit cooperative action over water resources. As the empirical studies have showed, water management institutions are often set up. Power asymmetry helps determine how these institutions operate. Shallow transboundary water interaction is a reflection of norm implementation being contested.

The three case studies exemplify this point. They indicate how applying common norms to guide river basin development projects was very problematic and time consuming. The Mahakali Treaty (1996), between Nepal and India, set out 'equal partnership' for project implementation, guided by cost and benefit sharing. However, applying such a norm to the implementation of the Pancheshwar Project was problematic because its treaty clauses were too general and did not specify the details of cost and benefit sharing. In the Lesotho–South Africa relationship the negotiations for the second phase of the LHWP exemplified shallow transboundary water interaction. According to the LHWP Treaty, a new agreement had to be reached for any additional phases of the project. This meant that treaty norms had to be translated into phase-specific agreements. South African decision-makers were initially non-committal about a second phase, arguing that all optimal plans for water supply – international and domestic – had to be thoroughly investigated. Negotiations over the second phase eventually commenced in 2004, much later than planned, and an initial agreement was reached only in 2008. In the Thailand–Vietnam case study, establishing legally binding principles for water utilization was a thorny issue from the time the multilateral river basin institution was set up. Although the MC discussed water utilization rules in the 1970s, no definitive binding document was produced. The establishment of a new quadrilateral committee in 1995, the MRC, provided a window of opportunity to revisit contentious issues. However, the outcome again yielded procedures with no legally binding powers.

Here we can see examples of the less-than-pretty face of cooperation. Zeitoun and Mirumachi (2008) argued that, depending on the extent to which it achieves equity, cooperation can have a pretty or an ugly face. In other words, cooperation may have a 'pretty' face when it contributes to equity in water resources management but be 'less pretty' when it does not effectively address inequalities and further institutionalizes them. The challenge of norm implementation exemplifies the situation where discursive power enables the hydro-hegemon to engage in cooperative water management institutions. However, at the same time, such power suppresses institutional development that could negatively influence its control over water allocation and river basin development. The bargaining tactics of the Thai hydrocracy, used to refute moves by other states trying to restrict water resources use, is one example of this. Similarly, the sanctioned discourse imposed upon discussions of the JCWR between Nepal and India restricted arrangements of water allocation to be fundamentally challenged and revisited. In addition, active stalling by South Africa, on decisions regarding the second phase of LHWP, is another example of an effective strategy employed by the hydro-hegemon under the guise of 'cooperation'. The effective use of soft power is particularly evident in the less pretty face of cooperation in transboundary water interactions (Zeitoun *et al.* 2011).

It is important that attention is paid to the use of power to sustain cooperation that benefits the hydro-hegemon. This is because the non-hydro-hegemons often have no alternative but to participate in shallow transboundary water interactions. For example, in the Ganges case study, upstream Nepal had little option but to engage with downstream India if it were to take advantage of hydropower potential: India was the only realistic market for hydropower export. A riparian position in this case does not provide much advantage for Nepal. This bilateral relationship was characterized by shallow transboundary water interaction much more so than the other two cases. While several bilateral projects, agreements and joint committees were set up, the elite decision-makers of Nepal were unable to change the rules of the game. As pointed out in Chapter 4 (this volume), the path dependency of river basin development emphasized the economic value of water resources. Such path dependency and power asymmetry is a potent mix. Agendas for water resources management become defined in very narrow terms, making it difficult for non-hydro-hegemons to suggest alternative rationales for collective action.

Power asymmetry is not unique to transboundary water politics. Certainly many aspects of socio-political life in general exemplify how power is used in both overt and subtle ways. The particular interest of power asymmetry in transboundary water politics is in thinking about the short- and long-term implications, and the accountability of such. As the case studies have showed, the short-term implications of power asymmetry are seen in the pace and progress of negotiations. Bilateral decision-making can be protracted, as in the case of the LHWP between South Africa and Lesotho over Phase 2, making negotiations a costly affair. The lack of project details for the Pancheshwar Project between Nepal and India is another short-term implication. It may be argued that the shallow transboundary water interaction between Thailand and Vietnam over rules and procedures of water use has, in part, contributed to the ineffectiveness of the MRC to deal with the Xayaburi dam controversy. In the long run, shallow transboundary water interaction influenced by power asymmetry also has the effect of justifying the hydraulic mission, even if there are repeated attempts at integrated, reflexive water resources management, a point to which we will return later.

Elite decision-making and problem frames

As mentioned earlier, highlighting the role of the hydrocracy and associated decision-makers provides the base of a critique on the way in which transboundary water resources are developed, managed and governed. With the TWINS approach, the focus on hydrocracies and politicians, as interlocutors of speech acts, enabled unpacking the 'national' interests related to international water resources management. In other words, national interests represent the views of a particular set of elite decision-makers. These interests are defined to maximize opportunities for economic development, whereby water resources are constructed as a utilitarian asset. Consequently, water resources and rivers are subject to commodification, either in their natural state (as in the case of the LHWP as water supply in urban areas) or through food and energy production (as in the case of projects along the Mekong and Ganges rivers).

This way of viewing the river basin and shared water resources by elite decision-makers may be explained further by considering uncertainty and problem frames. Uncertainty may be normative, about the fundamental plurality of views, which examine a problem and its solution (Newig *et al.* 2005; Brugnach *et al.* 2008). It may also be informational in nature, relating to imprecise, inaccurate or limited knowledge (Brugnach *et al.* 2008). The combination of normative and information uncertainty can frame river basin development as structured, moderately structured or unstructured problems (Hisschemöller and Hoppe 2001). Figure 7.2 shows how structured problems, with a low degree of

		Normative uncertainty	
		High	Low
Informational uncertainty	High	Unstructured problem with widespread recognition that the status quo needs to be addressed. Requires complex solutions.	Moderately structured problem requiring resolution as to the means of addressing the problem.
	Low	Moderately structured problem requiring a process that accommodates multiple views without alienation. Aimed at future consensus rather than immediate resolution.	Structured problem addressed with existing techniques and expertise.

Figure 7.2 Problem frames based on information and normative uncertainty (source: author, based on Hisschemöller and Hoppe (2001: 43–49)).

normative and information uncertainty, favour the application of existing, standardized measures based on professional or technical expertise. A moderately structured problem, with a low degree of normative uncertainty but high informational uncertainty, requires consideration of how the problem may be solved. Conversely, a moderately structured problem, with a high degree of normative uncertainty and low informational uncertainty, cannot be solved immediately. Instead it requires consideration of multiple perspectives in a process aimed at future consensus. The result is a 'freezing' or depoliticization of the issue at stake (ibid.: 48–49). An unstructured problem, with both high informational and normative uncertainty, is hard to define, and thus results in a deeply political process of problem structuring (ibid.: 51). This type of problem calls for more non-linear, interdisciplinary approaches and for the inclusion of a wider range of stakeholders. Decision-makers or technocrats alone cannot solve this kind of problem in the same way as a structured one.

Water issues concerning transboundary water interactions observed in the three case studies tended to be framed as either structured or moderately structured problems by the elite decision-makers. The perceived problem was that the river did not flow in a way that benefited society. The solution would be to intervene with dams, irrigation canals and transfer tunnels. The hydraulic mission paradigm (Allan 2001) presided over the ways in which water resources should be managed, privileging only one type of rationality. Decision-making by elites, firmly based in the ideals and goals of the hydraulic mission, leaves little room for alternative ideas about how water resources could be developed. In other words, the hydrocracy did not perceive high normative uncertainty of river basin management. The exclusive nature of decision-making taken to achieve the hydraulic mission does not reflect such plurality. Rather, data and information are seen as important inputs to better decision-making. For example, the utilitarian use of cost–benefit analysis relates to reducing informational uncertainty. The complexities of the impact of river basin development are conjugated to measurable issues that may be monitored and evaluated.

Developing transboundary water resources tends to focus on addressing informational uncertainty as the main focus when viewed from the elite decision-makers. The implication is that river basin development is portrayed as a largely *technical* issue where equality of costs and benefits needs to be negotiated among states. Structured problem frames in particular further masculinize transboundary water resources management. Sultana (2004) pointed out that international politics among the basin states of the Ganges River basin are structured around masculine discourses about river engineering. She argued strongly that any major engineering project to divert river flow between the basin states is bound to be problematic. This is because the link between decisions at the basin level and their local implications are obscured and ignored. Engineering intervention exacerbates the local challenges women face in ensuring water quality and quantity, and ultimately in dealing with poverty. If regional cooperation is to be meaningful, gendered considerations of river basin development need to be incorporated (ibid.).

The TWINS analysis showed that this combination of problem frames and solution is held by hydrocracies of both the hydro–hegemons and non–hydro–hegemons. Equity of water allocation was contested but the *purpose* of developing the river basin was not. While common norm implementation was challenging, the case studies showed that there was consensus on common goals among the key decision-makers, made up of hydrocracies and politicians. In the three case studies, the establishment of common goals occurred relatively early on in the respective bilateral relationships. In the Nepal–India relationship, the common goal was to develop multi-purpose hydraulic infrastructure on the tributaries of the Ganges River. The Kosi, Gandak, Karnali and Pancheshwar projects were negotiated on the basis of this common goal. Despite the deadlock over the Mahakali Treaty between Nepal and India, water for economic development was a clear imperative. In the Lesotho–South Africa relationship, the common goal was to transfer water from Lesotho to South Africa by constructing hydraulic infrastructure. The development of an upstream bloc between South Africa and Lesotho in ORASE-COM negotiations was an attempt to buffer any negative impact upon the established bilateral water allocation agreement. Even though there had been some historically contentious transboundary water interactions, the rationality of securing water to reap economic benefits was shared by the elite decision-makers of South Africa and Lesotho. In the Thailand–Vietnam relationship, the common goal was to develop the mainstream and tributaries of the Mekong River, largely to meet irrigation and hydropower demands. The intensified interest in hydropower opportunities in the Mekong River basin is particularly illustrative, as decision-makers from both hydro-hegemonic and non-hydro-hegemonic states saw great potential in dam development. Even though Laos had not driven the agenda of the RBOs in the past, its interest in capitalizing on water resources for hydropower development helped form a unified rationality with Thai decision-makers. Another aspect of this hydropower development is that dam building as a solution is also strengthened by the participation of the private sector that mobilizes financing. The shared problem framing of water resources management and the solutions to it were projected on to transboundary water interaction. Maximizing the economic utility of shared water resources is uniform and compatible across states between the hydrocracies.

This point about a specific rationality shared by both the hydro-hegemon and non-hydro-hegemons necessitates further thinking of equity. In FHH, as a result of treating the state as a given, power asymmetry is considered to hinder equity between states (Zeitoun and Warner 2006). However, if the hydrocracies in both the hydro-hegemonic and non-hydro-hegemonic states have a shared rationality and benefit from it (though to differing extents), the real repercussions are on sub-national stakeholders exempt from this opaque decision-making process. The problem is not that the river basin is the focus of development. Rather it is what the hydraulic mission, charged with power asymmetry, ends up marginalizing and devaluing: alternative worldviews of development; costs and burdens to not only the less powerful state, but also to local communities across and within borders of the hydro-hegemon; impacts to ecosystems and environmental health; and cultural and social values related to water resources not easily expressed in economic terms. Moreover, these trade-offs can have a long-term effect; much longer than the life span of a dam. Understanding shallow transboundary water interaction points to problems of elitist, gendered decision-making that supports the hydraulic mission and encourages the commodification of water resources. The actual costs and benefits to local communities and individuals relying on the river basin are not necessarily clear, and, at worst, are only considered through the direct gains afforded to the elite decision-makers. These situations are what Blatter and Ingram (2000: 446–447) described as decision-making 'under a cloak of diplomatic secrecy, which limits the opportunities for local citizens – exposed to contamination and water shortages – to either understand or act to ameliorate these problems', as in their study of transboundary water issues between the USA and Mexico.

These problem frames and rationalities, underpinned by the hydraulic mission, reflect the way in which the hydrocracies actively socialize nature (Castree 2001). Through feasiblity studies and investigations, the hydrocracies of basin states each developed their position regarding the optimization of water resources that were then negotiated at the international level. In doing so, technical knowledge of the river and how to control it is privileged, discounting alternative forms of knowledge (Swatuk and Wirkus 2009). Cost–benefit analysis provides huge scope for criticism, as non-material costs are not rigorously considered (Mirumachi and Torriti 2012). A narrow understanding of knowledge also throws up problems of how to deal with impacts from hydraulic development. There are numerous examples where large-scale dam projects have failed to take into consideration environmental and socioeconomic impacts. Even if these factors are considered, mitigation and adaptation measures are poorly devised (Scudder 2005). It has been suggested that the larger the scale of hydraulic infrastructure for storage, the stronger institutional capacity needs to be for equity and to address negative impacts (van der Zaag and Gupta 2008). It would seem that, based on the structured problem approach taken by the hydrocracy, there would be limited institutional capacity to deal with the intangible, hard-to-quantify socio-political impacts that are outside of the technical mindset.

Resilience of the hydrocracy

Understanding problem frames can complement the explanation of how subnational decision-making and international transboundary water negotiations are linked. National interests presented by the hydrocracies at international negotiations are narrowly defined versions of water use for economic development. Such interests are not necessarily embraced by all stakeholders, and the allocation of water resources, as well as the costs and benefits from developing the river, is certainly contentious. The case studies provide a snapshot of such contentions. For example, civil society concerns about the LHWP in its failure to deal with major socio-economic and environmental impacts reveal the fundamental differences of rationality among these groups of stakeholders. In the Mekong River basin, while the degree of maturity of civil society differs among basin states, there is mounting concern about the pace at which hydro-power development is occurring. Civil society campaigns have opened up discussions on whether dams should be built or not, questioning what a 'good' dam might entail – a deeply normatively uncertain issue. In Nepal, the Kosi and Gandak projects were met with fierce discontent from both political parties and the general public, and the Tanakpur project spurred major political debate. The case of Nepal shows how environmental issues can rise to the tops of national agendas and challenge governmental decisions. However, the fundamental question on how the actual water users would benefit from the projects was under-examined, and issues of sovereignty and independence from India took centre stage in many of these political debates. Alternative voices, critical of large-scale projects, were simply not taken seriously enough (see Gyawali 2001).

Public participation has now become one of the key features of water governance. Best represented in efforts to put in place IWRM, public participation is regarded as a necessary component to manage water resources. IWRM at the transboundary level is occurring, though at an uneven pace and scope across various basins (Hooper and Lloyd 2011). However, RBOs like the MRC have taken steps to institutionalize public participation. During the PNPCA process for the Xayaburi dam, the national committees organized public meetings to deliberate the manner. These meetings are designed to feed into the decisions communicated via the NMCs to the MRC. In the Southern African case study, public participation has also been an important element in developing the water governance structure at the regional and basin-wide levels. However, the path dependency of the hydraulic mission, stemming much earlier than when IWRM became mainstreamed in the 1990s, is not negligible. As explained in Chapter 4 (this volume), the path dependency of developing water resources through specific projects entrenches interests of the elite decision-makers in material ways.[1]

The elite decision-makers hold beliefs of high modernity that reflect progress through rational and scientific approaches to controlling nature (Baghel and Nüsser 2010: 237). Dams were the pinnacle of such ideals (ibid.; Kaika 2006). However, it should not be forgotten that, in the case of large hydro-power projects, their opportunity costs prohibit alternatives for other modes of energy production (Ansar *et al.* 2014). Mega-projects ultimately make little economic sense because of their high costs that overrun initial budgets (Flyvberg 2014). None the less, these projects are highly favoured. Elite decision-makers hold 'the conviction that it is an important duty of the state to develop water resources' (Wester 2008: 10). As Hori (1996: 164; my translation) reflected about plans for development in the Mekong River basin:

With hindsight, it seems amazingly optimistic but in the world during the 1960s, there were none who would doubt the need and possibility of such development. It was indeed expected that any major development project would be possible, as long as funds could be raised.

Consequently, the role of the state, mediated through elite decision-makers, has a strong history and dominance.

The discourses presented by these elite decision-makers thus tend to be resilient to alternative discourses. The hydrocracies reflect the power of a state by implementing highly prioritized and iconic projects of national development (Molle *et al.* 2009). However, there are other reasons than a historical role that make them particularly resilient. Hydrocracies tend to be organizations that can deploy strategies to ensure organizational survival. Molle and colleagues (2009: 341) claimed that hydrocracies reorganize themselves superficially 'under the guise of apparently drastic institutional reforms', such as reflexive water management policy. None the less, they also maintain rationalities of the hydraulic mission. In the Ganges River basin and the greater GBM basin, the hydrocracy is characterized as resilient to change and challenges to the engineered, large-scale approach of water management (Bandyopadhyay and Ghosh 2009: 54). In Southern Africa, the hydrocracies benefit from the popular belief that 'the hard path to water development ... facilitate[s] economic development and so deliver[s] jobs, votes, money, influence and power' (Swatuk 2008: 41). While policy language over transboundary water management may focus on development and sustainability, large-scale development remains on the national and bilateral agendas.

The resilience of the hydrocracy does not necessarily mean that there are no competing views within the hydrocracy. As Swatuk (2008) and Wester (2008) found, the hydrocracy is usually complex, and comprises various individuals and organizations. Thus, 'the hydrocracy are not monolithic and how strong or contested they are in different time periods and countries is an empirical question' (Wester 2008: 10). For example, ministries involved in water resources management have competed against each other to assert their organizational legitimacy in Vietnam (Molle and Hoanh 2009).[2] When these bureaucratic struggles are reflected in RBOs, they have the effect of weakening the use of the multilateral fora for basin-wide governance. Suhardiman and colleagues (2012) described this as the 'scalar disconnect' between national agendas and policies, and decisions in the MRC. The hydrocracies can still exert their influence outside of the RBO, as the example of hydropower development by the Thai hydrocracy showed in Chapter 6 (this volume). The different fora that the hydrocracy can use to determine both domestic and international transboundary waters hint at how further research can incorporate notions of governmentality, or the ways in which state power is exercised, through knowledge and techniques (Foucault 1991). Analysing the specific ways in which hydrological modelling and monitoring of the river system, cost–benefit analysis of water use or engineering expertise of large-scale infrastructure is deployed, and also challenged by civil society and other organizations vying for state power, could be useful.

Practising transboundary water governance

Each of the case studies explored the implications of how transboundary water interactions evolved, focusing in particular on basin-wide water governance. In the Ganges River basin, path dependency entrenches and prioritizes bilateral agreements rather than basin-wide arrangements for their supposed efficiency. In the case of the Orange–Senqu River basin, 'regional interests' are not a given, but rather are contested by upstream and downstream basin states, despite the number of water governance institutions put in place. These contestations stem from bilateral water arrangements separate from multilateral water governance initiatives. In the Mekong River basin, the notion of the 'Mekong spirit' enabled a mode of governance that is seemingly cooperative but which serves the particular interests of individual basin states, rather than the basin as a whole.

Multilateral water governance has been promoted, based on the argument that it provides a 'win-win' situation where basin states would benefit, regardless of their natural endowment (Sadoff and Grey 2002, 2005). In theory, basin-wide transboundary water management would provide a regional public good (Nicol *et al.* 2001). However, there are some more nuanced views, qualifying when multilateral institutions would be required and necessary. For example, in discussing the international architecture of transboundary water management, the DFID and WWF (2010) specified that agreements need to be selectively developed because there are basins that do not necessarily require the further formalization of rules. Water governance at the transboundary level is often synonymous with RBOs, as the vehicle for multilateral cooperation and rule-making. However, insightful arguments about the scope of RBOs are made by Jensen and Lange (2013: 115), who claim that 'RBOs are still too water-centric, meaning that they are not engaging effectively with ministries that are in a position to take important water governance issues to a political level, such as ministries of foreign affairs, energy and finance'. Some further question the fundamental use of RBOs and the mode of governance it pursues. For example, Merrey (2009) provided a very harsh critique of RBOs that are being promoted and funded by Western donors in African transboundary river basins. He argued that these RBOs, despite promoting public participation, actually do very little of that and this has the effect of dismissing the local knowledge and practice of water resources management. A 'Western model' of governance imposed upon an incongruent context is also pointed out by Hensengerth (2009) in the study of the Mekong River basin. These are useful critiques in that they begin to highlight the challenges of accounting for river basin governance.

Policy for transboundary water governance needs first to critically revisit what value to the regional public good RBOs bring about and to whom.

Challenging problem frames through bilateral and multilateral governance initiatives is necessary but needs to be planned with a long-term vision. Co-learning among different stakeholders, or social learning, is seen as an important element in revitalizing governance in the water sector. However, as many studies note, this process is not without its uncertainty for success and requires time (e.g. Armitage *et al.* 2008; Lebel *et al.* 2010; Huntjens *et al.* 2012: 73). Accurately assessing power relations and the way in which trans-boundary water interactions have changed over time can contribute to addressing such challenges.

Second, accounting for effective outcomes of transboundary water govern-ance needs to be regarded as a matter beyond institutional mechanisms. Hydro-hegemony operating at different spatial scales, as the empirical case studies showed, enables water resources control through multiple projects and means. It has been reported that treaties regarding international transboundary river basins have developed over time to become comprehensive, fore-grounding environmental management as opposed to the regulation and abstraction of water resources (Giordano *et al.* 2014). Efforts at environmental conservation and stewardship are important and overdue in many parts of the world. However, closer scrutiny is required to understand how these treaties actually influence management practices on the ground as a result of being comprehensive. The multi-scalar nature of the reaches of transboundary water politics makes RBO decision-making challenging because the organization would need to have sufficient geographic coverage and not be limited to spe-cific projects. In effect, accounting for the transboundary implications of river basin development cannot be restricted to the functions of RBOs, which are one type of governance mechanism. RBO cannot replace state decision-making. National regulatory capacities become crucial to harmonize policies across scales (Mirumachi 2013). After examining wetland management in India, Narayanan and Venot (2009: 330) argued that functioning regulatory powers of the state that do not favour certain stakeholders at the expense of others is important. This is particularly so because the decisions made by gov-ernments can trigger detrimental effects upon both the environment and society. While their argument was made based on studies at the national level, this insight is also applicable to international transboundary river basin pro-jects. Seeking accountability through regulatory measures can also be one way to further deliberate the meaning of equity and how to realize them in par-ticular contexts.

While being power-blind is detrimental, policy for transboundary water governance cannot overlook the importance of improving the material capa-bility of individual basin states especially in developing regions. The financial and technical resources need to be developed in a way that would allow for a wider set of options for water resources management, and to shape, adapt and challenge water governance mechanisms that may already be in place (with the aid of donors). As the Mekong hydropower issues showed, the make-up of stakeholders is increasingly heterogeneous and forms networks across sectors and scales. Countering dominant problem frames and rationalities for transboundary water governance thus requires changes at both the national as well as the transboundary level.

Keeping transboundary water politics political

The fact that transboundary water management is political is not unique to other examples of environmental management and governance. The transboundary scale may increase complexity in the making and understanding of problems and solutions to shared waters. However, struggles for equity, challenging power asymmetries and seeking sustainable access and allocation are very fundamental elements to many environmental issues and they make up transboundary water politics. While this book has focused on developing country contexts, the politics of water are also evident in the Global North. Moreover, these social processes are ongoing, marking everyday practices of water resources management and manifesting in key milestones or crunch points of transboundary water interactions. Geographic imaginations are re-invented, and notions of basins are territorialized and deterritorialized through negotiations to maintain or challenge control over water resources. It is therefore important that transboundary water politics is kept political. Apoliticizing transboundary waters would obscure the ways in which decisions over river basin development have far-reaching implications for agriculture, energy, ecosystems, livelihoods, regional stability and human security. Developing and refining analytical approaches to transboundary water politics is needed, as well as policy engagement. The development of the TWINS approach and the elaboration of case studies represented one way to contribute to this purpose. Keeping transboundary water politics political requires the continuation of asking critical questions: who is involved in decision-making and who is not; what kind of power is being exerted with what means; how do transboundary water interactions change over time and with what effects.

Notes

1 See also Sehring (2009) for a Central Asian example of the way in which path dependency works.
2 Molle and Hoanh (2009) attributed donor intervention to promote IWRM as one of the reasons for this internal fragmentation.

References

2030 Water Resources Group 2009, *Charting Our Water Future: Economic Frameworks to Inform Decision-Making*, 2030 Water Resources Group. Available at: www.mckinsey.com/clientservice/water/charting_our_water_future.aspx.

Abukhater, A. 2013, *Water as a Catalyst for Peace: Transboundary Water Management and Conflict Resolution*, Routledge, London.

Adhikary, K.D., Ahmad, Q.K., Malla, S.K., Pradhan, B.B., Rahman, K., Rangachari, R., Rasheed, K.B.S. and Verghese, B.G. (eds) 2000, *Cooperation on the Eastern Himalayan Rivers: Opportunities and Challenges*, Konark Publishers, Delhi.

Agnew, J. 2011, 'Waterpower: Politics and the geography of water provision', *Annals of the Association of American Geographers*, vol. 101, no. 3, pp. 463–476.

Agreement, 1995, *Agreement on the Cooperation for the Sustainable Development of the Mekong River Basin*, Chiang Rai, Thailand, 5 April.

Akanda, A.S. 2012, 'South Asia's water conundrum: Hydroclimatic and geopolitical asymmetry, and brewing conflicts in the Eastern Himalayas', *International Journal of River Basin Management*, vol. 10, no. 4, pp. 307–315.

Al Jazeera 2014, 'Egypt to "escalate" Ethiopian dam dispute', Al Jazeera, 21 April. Available at: www.aljazeera.com/news/middleeast/2014/04/egypt-escalate-ethiopian-dam-dispute-201448135352769150.html.

Allan, J.A. 2001, *The Middle East Water Question: Hydropolitics and the Global Economy*, I.B. Tauris, London; New York.

—— 2002, 'Hydro-peace in the Middle East: Why no water wars? A case study of the Jordan River Basin', *SAIS Review*, vol. 12, no. 2, pp. 255–272.

—— 2003, *IWRM/IWRAM: A new sanctioned discourse?*, SOAS/KCL Water Issues Group Occasional Paper 50, SOAS/King's College London, London.

—— 2007, 'Political Economy of Power and Water: Introductory Orientation', Third International Workshop on Hydro-Hegemony, 12–13 May, London.

—— 2011, *Virtual Water: Tackling the Threat to our Planet's Most Precious Resource*, I.B. Tauris, London; New York.

—— 2013, 'Food-water security: Beyond water resources and the water sector', in *Water Security: Principles, Perspectives and Practices*, ed. B. Lankford, K. Bakker, M. Zeitoun and D. Conway, Routledge, London, pp. 321–335.

Allan, J.A. and Mirumachi, N. 2010, 'Why negotiate? Asymmetric endowments, asymmetric power and the invisible nexus of water, trade and power that brings apparent water security', in *Transboundary Water Management: Principles and Practice*, ed. A. Earle, A. Jägerskog and J. Öjendal, Earthscan, London; Washington, DC, pp. 13–26.

Allouche, J. 2005, *Water Nationalism: An Explanation of the Past and Present Conflicts in Central Asia, the Middle East and the Indian Subcontinent?*, Ph.D. thesis, Université de Genève, Geneva.

Ambrose, D. 1998, 'Text of Prime Minister's letter to SADC Heads of State', *Summary of Events in Lesotho*, vol. 5, no. 3. Available at: www.trc.org.ls/events/events19.983.htm.

Amended Gandak Agreement 1964, *Amended Agreement between His Majesty's Government of Nepal and the Government of India on the Gandak Irrigation and Power Project*, signed at Kathmandu on 30 April.

Amended Kosi Agreement 1966, *Amended Agreement between His Majesty's Government of Nepal and the Government of India concerning the Kosi Project*, signed at Katmandu on 19 December.

Ansar, A., Flyvbjerg, B., Budzier, A. and Lunn, D. 2014, 'Should we build more large dams? The actual costs of hydropower megaproject development', *Energy Policy*, vol. 69, pp. 43–56.

Archer, R. 1996, *Trust in Construction? The Lesotho Highlands Water Project*, Christian Aid, London.

Armitage, D., Marschke, M. and Plummer, R. 2008, 'Adaptive co-management and the paradox of learning', *Global Environmental Change*, vol. 18, no. 1, pp. 86–98.

Austin, J.L. 1962, *How to Do Things with Words*, Clarendon Press, Oxford.

Aviram, R., Katz, D. and Shmueli, D. 2014, "Desalination as a game-changer in transboundary hydro-politics', *Water Policy*, vol. 16, pp. 609–624.

Axelrod, R.M. 1984, *The Evolution of Cooperation*, Basic Books, New York.

Axelrod, R.M. and Keohane, R.O. 1985, 'Achieving cooperation under anarchy: Strategies and institutions', *World Politics*, vol. 38, no. 1, pp. 226–254.

Babel, M.S. and Wahid, S.M. 2008, *Freshwater Under Threat: South Asia – Vulnerability Assessment of Freshwater Resources to Environmental Change*, UNEP, Nairobi.

Baghel, R. and Nüsser, M. 2010, 'Discussing large dams in Asia after the World Commission on Dams: Is a political ecology approach the way forward?', *Water Alternatives*, vol. 3, pp. 231–248.

Baillat, A. 2010, *International Trade in Water Rights: The Next Step*, IWA Publishing, London.

Bakker, K. 1999, 'The politics of hydropower: Developing the Mekong', *Political Geography*, vol. 18, no. 2, pp. 209–232.

—— 2013, 'Constructing "public" water: The World Bank, urban water supply, and the biopolitics of development', *Environment and Planning D: Society and Space*, vol. 31, no. 2, pp. 280–300.

Bakker, M.H.N. 2009, "Transboundary river floods and institutional capacity', *Journal of the American Water Resources Association (JAWRA)*, vol. 45, no. 3, pp. 553–566.

Bandyopadhyay, J. and Ghosh, N. 2009, 'Holistic engineering and hydro-diplomacy in the Ganges–Brahmaputra–Meghna Basin', *Economic and Political Weekly*, vol. 44, no. 45, pp. 50–60.

Bangkok Post 1977, 'New hope for Mekong Project', *Bangkok Post*, 24 April, p. 1.

—— 1992, 'Thailand seeks Mekong delay', *Bangkok Post*, 19 February, p. 3.

Baran, E., Larinier, M., Ziv, G. and Maramulla, G. 2011, *Review of the Fish and Fisheries Aspects in the Feasibility Study and the Environmental Impact Assessment of the Proposed Xayaburi Dam on the Mekong Mainstream*. Report prepared for the WWF Greater Mekong. Available at: assets.panda.org/downloads/wwf_xayaburi_dam_review310311.pdf.

Barker, R., Meinzen-Dick, R., Shah, T., Tuong, T.P. and Levine, G. 2010, 'Managing irrigation in an environment of water scarcity', in *Rice in the Global Economy: Strategic Research and Policy Issues for Food Security*, ed. S. Pandey, D. Byerlee, D. Dawe, A. Dobermann, S. Mohanty, S. Rozelle and B. Hardy, International Rice Research Institute, Los Baños, pp. 265–296.

Barnett, J. 2000, 'Destabilizing the environment-conflict thesis', *Review of International Studies*, vol. 26, pp. 271–288.

—— 2001, *The Meaning of Environmental Security: Ecological Politics and Policy in the New Security Era*, Zed Books, London.

Barney, K. 2009, 'Laos and the making of a "relational" resource frontier', *Geographical Journal*, vol. 175, no. 2, pp. 146–159.

British Broadcasting Corporation (BBC) 1980, 'S Africa–Lesotho Hydroelectric and Irrigation Project', *BBC Summary of World Broadcasts*, 2 September.

—— 1983, East Africa: In brief; Lesotho Minister asks S Africa to prove terrorism allegations', *BBC Summary of World Broadcasts*, 28 May.

—— 1985, 'Thailand boycotts Hanoi Mekong committee meeting', *BBC Summary of World Broadcasts*, 15 January.

—— 1986a, 'Lesotho's Lekhanya on normalising relations with S Africa', *BBC Summary of World Broadcasts*, 1 March.

—— 1986b, 'South Africa and Lesotho', *BBC Summary of World Broadcasts*, 26 March.

—— 1986c, 'Meeting of Interim Mekong Committee in Thailand', *BBC Summary of World Broadcasts*, 26 July.

—— 1992, 'Vietnam denounces Thailand's "obstructive move" against Mekong Committee', *BBC Summary of World Broadcasts*, 9 March.

—— 1994, 'Water Resources Minister interviewed on Mekong River basin cooperation', *BBC Summary of World Broadcasts*, 6 January.

—— 2013, 'Egyptian warning over Ethiopia Nile dam', 10 June. Available at: www.bbc.co.uk/news/world-africa-22850124.

Berardo, R. and Gerlak, A. 2012, 'Conflict and cooperation along international rivers: Crafting a model of institutional effectiveness', *Global Environmental Politics*, vol. 12, no. 1, pp. 101–120.

Bernauer, T. 2002, 'Explaining success and failure in international river management', *Aquatic Sciences*, vol. 64, pp. 1–19.

Bernauer, T. and Kalbhenn, A. 2010, 'The politics of freshwater resources', in *The International Studies Encyclopedia*, ed. A. Denemark, Wiley-Blackwell, Oxford, pp. 5800–5821.

Bhadra, B. 2004, 'Hydropower development in Nepal: Problems and prospects', in *Nepalese Economy: Towards Building a Strong Economic Nation-state*, ed. M.K. Dahal, Central Department of Economics, Tribhuvan University, and New Hira Books Enterprises, Kathmandu.

Bhasin, A.S. (ed.) 2005, *Nepal–India, Nepal–China Relations: Documents 1947–June 2005, Volume 2*, Geetika Publishers, New Delhi.

Biggs, D., Miller, F., Hoanh, C.T. and Molle, F. 2009, 'The delta machine: Water management in the Vietnamese Mekong Delta in historical and contemporary perspectives', in *Contested Waterscapes in the Mekong Region: Hydropower, Livelihoods and Governance*, ed. F. Molle, T. Foran and M. Käkönen, Earthscan, London, pp. 203–225.

Biswas, A.K. 2011, 'Cooperation or conflict in transboundary water management: Case study of South Asia', *Hydrological Sciences Journal*, vol. 56, no. 4, pp. 662–670.

Blatter, J. and Ingram, H. 2000, 'States, markets and beyond: Governance of trans-boundary water resources', *Natural Resources Journal*, vol. 40, no. 2, pp. 439–473.

Boadu, F.O. 1991, 'Law and natural-resource management: Case of water in Lesotho', *Journal of Water Resources Planning and Management*, vol. 117, no. 6, pp. 698–710.

—— 1998, 'Relational characteristics of transboundary water treaties: Lesotho's water transfer treaty with the Republic of South Africa', *Natural Resources Journal*, vol. 38, no. 3, pp. 381–410.

Braun, D. 1986, 'Two weeks and Maseru toes line', *The Star*, 15 January, p. 4.

Brichieri-Colombi, S. and Bradnock, R.W. 2003, 'Geopolitics, water and development in South Asia: Cooperative development in the Ganges-Brahmaputra delta', *The Geographical Journal*, vol. 169, no. 1, pp. 43–64.

Brochmann, M. and Hensel, P.R. 2009, 'Peaceful management of international river claims', *International Negotiation*, vol. 14, no. 2, pp. 393–418.

—— 2011, 'The effectiveness of negotiations over international river claims', *International Studies Quarterly*, vol. 55, pp. 1–24.

Browder, G. 1998, *Negotiating an International Regime for Water Allocation in the Mekong River Basin*, Ph.D. thesis, Stanford University, Stanford, CA.

Browder, G. and Ortolano, L., 2000, 'The evolution of an international water resources management regime in the Mekong River basin', *Natural Resources Journal*, vol. 40, no. 3, pp. 499–531.

Brugnach, M. and Ingram, H. 2012, 'Ambiguity: The challenge of knowing and deciding together', *Environmental Science and Policy*, vol. 15, no. 1, pp. 60–71.

Brugnach, M., Dewulf, A., Pahl-Wostl, C. and Taillieu, T. 2008, 'Toward a relational concept of uncertainty: About knowing too little, knowing too differently, and accepting not to know', *Ecology and Society*, vol. 13, no. 2, p. 30.

Buzan, B., Wæver, O. and de Wilde, J. 1998, *Security: A New Framework for Analysis*, Lynne Rienner, Boulder, CO.

Cascão, A.E. 2008, 'Ethiopia – Challenges to Egyptian hegemony in the Nile Basin', *Water Policy*, vol. 10, no. S2, pp. 13–28.

—— 2009a, 'Changing power relations in the Nile River Basin: Unilateralism vs. cooperation?', *Water Alternatives*, vol. 2, no. 2, pp. 245–268.

—— 2009b, *Political Economy of Water Resources Management and Allocation in the Eastern Nile River Basin*, unpublished Ph.D. thesis, King's College London, London.

Castree, N. 2001, 'Socializing nature: Theory, practice, and politics', in *Social Nature: Theory, Practice and Politics*, ed. N. Castree and B. Braun, Blackwell, Malden, MA, pp. 1–21.

Central Planning and Development Office (Government of Lesotho) 1977, *Kingdom of Lesotho: Donor Conference Papers September 1977*, Government of Lesotho, Maseru.

Chakraborty, R. and Serageldin, I. 2004, 'Sharing of river waters among India and its neighbors in the 21st century: War or peace?', *Water International*, vol. 29, no. 2, pp. 201–208.

Checkel, J.T. 1999, 'Social construction and integration', *Journal of European Public Policy*, vol. 6, no. 4, pp. 545–560.

Chellaney, B. 2013, *Water: Asia's New Battleground*, Georgetown University Press.

Chenoweth, J. 2008, 'A re-assessment of indicators of national water scarcity', *Water International*, vol. 33, no. 1, pp. 5–18.

Chi, B.K., 1997, *From Committee to Commission? The Evolution of the Mekong River Agreements*, Ph.D. thesis, University of Melbourne.

Cleaver, F. and Elson, D. 1995, *Women and Water Resources: Continued Marginalisation and New Policies*, IIED, London.

Collins, D.N., Davenport, J.L. and Stoffel, M. 2013, 'Climatic variation and runoff from partially-glacierised Himalayan tributary basins of the Ganges', *Science of the Total Environment*, vols 468–469, Supplement, pp. S48–S59.

Conca, K. 2006, *Governing Water: Contentious Transnational Politics and Global Institution Building*, MIT Press, Cambridge, MA; London.

Conca, K., Wu, F. and Mei, C. 2006, 'Global regime formation or complex institution building? The principled content of international river agreements', *International Studies Quarterly*, vol. 50, no. 2, pp. 263–285.

Conti, K.I. 2014, *Enabling Factors for Transboundary Aquifer Cooperation: A Global Analysis*, International Groundwater Resources Assessment Centre (IGRAC), Delft.

Cosens, B.A. and Williams, M.K. 2012, 'Resilience and water governance: Adaptive governance in the Columbia River Basin', *Ecology and Society*, vol. 17, no. 4, p. 3.

Cowell, A. 1986, 'Military topples Lesotho leader; Capital jubilant', *New York Times*, 21 January, p. 3.

Craig, J.G. 1993, *The Nature of Co-operation*, Black Rose Books, Montréal; New York.

Crow, B. and Singh, N. 2000, 'Impediments and innovation in international rivers: The waters of South Asia', *World Development*, vol. 28, no. 11, pp. 1907–1925.

—— 2009, 'The management of international rivers as demands grow and supplies tighten: India, China, Nepal, Pakistan, Bangladesh', *India Review*, vol. 8, no. 3, pp. 306–339.

Crow, B. and Sultana, F. 2002, 'Gender, class, and access to water: Three cases in a poor and crowded delta', *Society and Natural Resources*, vol. 15, no. 8, pp. 709–724.

Dalby, S. 1992, 'Ecopolitical discourse: "Environmental security" and political geography', *Progress in Human Geography*, vol. 16, no. 4, pp. 503–522.

—— 2009, *Security and Environmental Change*, Polity Press, Cambridge.

Daoudy, M. 2005, *Le Partage des Eaux Entre la Syrie, l'Irak et la Turquie: Négociation, Sécurité et Asymétrie des Pouvoirs*, CNRS Editions, Paris.

—— 2009, 'Asymmetric power: Negotiating water in the Euphrates and Tigris', *International Negotiation*, vol. 14, no. 2, pp. 361–391.

Davidsen, P.A. 2006, *The Making and Unmaking of the Politics of Exceptionality: Studying Processes of Securitisation and Desecuritisation in the Orange and Okavango River Basins*, Master's thesis, Institute of Comparative Politics, The University of Bergen, Bergen.

De Stefano, L., Edwards, P., de Silva, L. and Wolf, A.T. 2010, "Tracking cooperation and conflict in international basins: Historic and recent trends", *Water Policy*, vol. 12, no. 6, pp. 871–884.

De Stefano, L., Duncan, J., Dinar, S., Stahl, K., Strzepek, K.M. and Wolf, A.T. 2012, 'Climate change and the institutional resilience of international river basins', *Journal of Peace Research*, vol. 49, no. 1, pp. 193–209.

Declaration 1978, *Declaration Concerning the Interim Committee for Coordination of Investigations of the Lower Mekong Basin*, signed by the representatives of the Governments of Laos, Thailand and Vietnam to the Committee for Coordination of Investigations of the Lower Mekong Basin at Vientiane on 5 January.

Department of Information (Government of South Africa) 1971, *Taming a River*

Giant: Story of South Africa's Orange River Project, Department of Information, Pretoria.

Department of Water Affairs (Government of Botswana) 2013, *Botswana Integrated Water Resources Management and Water Efficiency Plan*, DWA, Gaborone.

Detraz, N. 2009, 'Environmental security and gender: Necessary shifts in an evolving debate', *Security Studies*, vol. 18, no. 2, pp. 345–369.

Deudney, D. 1990, 'The case against linking environmental degradation and national security', *Millennium – Journal of International Studies*, vol. 19, no. 3, pp. 461–476.

DFID (Department for International Development, Government of UK) and WWF (World Wide Fund for Nature) 2010, *International Architecture for Transboundary Water Resources Management: Policy Analysis and Recommendations*, DFID, WWF, London.

Dhungel, D.N. 2009, 'Historical eye view', in *The Nepal–India Water Relationship: Challenges*, ed. D.N. Dhungel and S.B. Pun, Springer, Dordrecht, pp. 11–68.

Dhungel, D.N. and Pun, S. (eds) 2009, *The Nepal–India Water Relationship: Challenges*, Springer, Dordrecht.

Dhungel, K.R. 2004, *Readings in Nepalese Economy*, Adroit, New Delhi.

Dinar, S. 2009, 'Power asymmetry and negotiations in international river basins', *International Negotiation*, vol. 14, no. 2, pp. 329–360.

Dinar, S. and Dinar, A. 2000, 'Negotiating in international watercourses: Diplomacy, conflict and cooperation', *International Negotiation*, vol. 5, pp. 193–200.

Dixit, A. 1997, 'Indo–Nepal water resources development: Cursing the past or moving forward', in *India–Nepal Cooperation: Broadening Measures*, ed. J.R. Kumar, Department of History, University of Calcutta Monograph 13, K.P. Bagchi & Company, Calcutta, pp. 151–185.

Dixit, A., Adhikari, P. and Thapa, R. 2004, 'Nepal: Ground realities for Himalayan water management', in *Disputes over the Ganga*, ed. B. Subba and K. Pradhan, Panos Institute South Asia, Kathmandu, pp. 158–191.

Dombrowsky, I. 2007, *Conflict, Cooperation and Institutions in International Water Management: An Economic Analysis*, Edward Elgar, Cheltenham; Northampton, MA.

Dore, J. and Lazarus, K. 2009, 'De-marginalizing the Mekong River Commission', in *Contested Waterscapes in the Mekong Region: Hydropower, Livelihoods and Governance*, ed. F. Molle, T. Foran and M. Käkönen, Earthscan, London pp. 357–382.

Dore, J., Lebel, L. and Molle, F. 2012, 'A framework for analysing transboundary water governance complexes, illustrated in the Mekong Region', *Journal of Hydrology*, vols 466–467, pp. 23–36.

Duffy, G. and Frederking, B. 2009, 'Changing the rules: A speech act analysis of the end of the Cold War', *International Studies Quarterly*, vol. 53, no. 2, pp. 325–347.

Department of Water Affairs (DWA) (Government of South Africa) 1986, *Report of the Director-General: Water Affairs for the Period 1 April 1984 to 31 March 1985*, DWA, Pretoria.

—— 1987, *Water: Meeting the Needs of the Nation 1912–1987*, Department of Water Affairs, Pretoria.

—— 1989, *Report of the Director-General: Water Affairs for the Period 1 April 1987 to 31 March 1988*, DWA, Pretoria.

Department of Water Affairs and Forestry (DWAF) (Government of South Africa) 1997, *White Paper on a National Water Policy for South Africa*, DWAF, Pretoria.

—— 2000, *Thukela Water Project Feasibility Study. Instream Flow Requirement Study Summary Report*, prepared by Delana Louw for IWR Environmental as part of the

Thukela Water Project Feasibility Study, DWAF Report No. PBV000–00–7599, DWAF, Pretoria.

—— 2005, *Lesotho Highlands Water Project (LHWP)*, DWAF. Available at: www.info. gov.za/speeches/2005/05092712151006.htm.

—— 2001, *Thukela Water Project Feasibility Study, Main Feasibility Report*, prepared by BKS Incorporated Water Division as part of the Thukela Water Project Feasibility Study, DWAF Report No. PBV000–00–9700, DWAF, Pretoria.

—— 2014a, *Orange River Project*. Available at: www.dwaf.gov.za/orange/default.htm.

—— 2014b, *Thukela Water Project*. Available at: www.dwaf.gov.za/thukela/.

Earle, A. 2007, 'The role of governance in countering corruption: An African case study', *Water Policy*, vol. 9, Supplement 2, pp. 69–81.

Earle, A. and Bazilli, S. 2013, 'A gendered critique of transboundary water management', *Feminist Review*, vol. 103, no. 1, pp. 99–119.

Earle, A., Malzbender, D., Turton, A.R. and Manzungu, E. 2005, *A Preliminary Basin Profile of the Orange/Senqu River*, African Water Issues Research Unit (AWIRU), University of Pretoria, Cape Town.

Ebinger, C.K. 2011, *Energy and Security in South Asia: Cooperation Or Conflict?*, Brookings Institution Press, Washington, DC.

Ekers, M. and Loftus, A. 2008, 'The power of water: Developing dialogues between Foucault and Gramsci', *Environment and Planning D: Society and Space*, vol. 26, no. 4, pp. 698–718.

Electricity Generating Authority of Thailand (EGAT) 2011, *Annual Report 2011*, EGAT, Bangkok.

Elhance, A.P. 1999, *Hydropolitics in the Third World: Conflict and Cooperation in International River Basins*, United States Institute of Peace Press, Washington, DC.

Evers, H-D. and Benedikter, S. 2009, 'Hydraulic bureaucracy in a modern hydraulic society – Strategic group formation in the Mekong delta, Vietnam', *Water Alternatives*, vol. 2, no. 3, pp. 416–439.

Falkenmark, M. 1989, 'The massive water scarcity now threatening Africa: Why isn't it being addressed?', *Ambio*, vol. 18, no. 2, pp. 112–118.

—— 1990, 'Rapid population growth and water scarcity: The predicament of tomorrow's Africa', *Population and Development Review*, vol. 16, pp. 81–94.

Falkenmark, M., Lundqvist, J. and Widstrand, C. 1989, 'Macro-scale water scarcity requires micro-scale approaches', *Natural Resources Forum*, vol. 13, no. 4, pp. 258–267.

Finnemore, M. 1996, *National Interests in International Society*, Cornell University Press, Ithaca, NY.

Flyvberg, B. 2014, 'What you should know about megaprojects and why: An overview', *Project Management Journal*, vol. 45, no. 2, pp. 6–19.

Food and Agricultre Organization of the United Nations (FAO) 2012, *Irrigation in Southern and Eastern Asia in Figures: AQUASTAT Survey – 2011*, FAO, Rome.

Foucault, M. 1991, 'Governmentality', in *The Foucault Effect: Studies in Governmentality*, ed. G. Burchell, C. Gordon and P. Miller, University of Chicago Press, Chicago.

Frederking, B. 2003, 'Constructing post-Cold War collective security', *The American Political Science Review*, vol. 97, no. 3, pp. 363–378.

Freimond, C. and Pitso, M. 1984, 'Lesotho PM reacts to pact threats by SA', *Rand Daily Mail*, 31 August, p. 4.

Frey, F.W. 1993, 'The political context of conflict and cooperation over international river basins', *Water International*, vol. 18, pp. 54–68.

Frey, F.W. and Naff, T. 1985, 'Water: An emerging issue in the Middle East?', *Annals of the American Academy of Political and Social Science*, vol. 482, pp. 65–84.

FT Energy Newsletters 1996, 'Mahakali, Basin Accord', *FT Energy Newsletters*, 5 February, p. 11.

Furlong, K. 2006, 'Hidden theories, troubled waters: International relations, the "territorial trap", and the Southern African Development Community's transboundary waters', *Political Geography*, vol. 25, pp. 438–458.

Gandak Agreement 1959, *Agreement between His Majesty's Government of Nepal and the Government of India on the Gandak Irrigation and Power Project,* signed at Kathmandu on 4 December.

Garver, J.W. 1991, 'China–India rivalry in Nepal: The clash over Chinese arms sales', *Asian Survey*, vol. 31, no. 10, pp. 956–975.

Geldenhuys, D. 1982, 'South Africa's regional policy', in *Changing Realities in Southern Africa: Implication for American Policy*, ed. M. Clough, Institute of International Studies, University of California, Berkeley.

George, W.L. 2013, 'Ethiopia's plan to dam the Nile has Egypt fuming', *Time*, 28 June. Available at: http://world.time.com/2013/06/28/ethiopias-plan-to-dam-the-nile-has-egypt-fuming/.

Gerlak, A.K. and Grant, K.A. 2009, 'The correlates of cooperative institutions for international rivers', in *Mapping the New World Order*, ed. T.J. Volgy, Z. Šabi , P. Roter and A.K. Gerlak, Wiley-Blackwell, Malden, MA, pp. 114–147.

Gilmont, M. 2014, 'Decoupling dependence on natural water: Reflexivity in the regulation and allocation of water in Israel', *Water Policy*, vol. 16, no. 1, pp. 79–101.

Giordano, M.A. and Wolf, A.T. 2003, 'Sharing waters: Post-Rio international water management', *Natural Resources Forum*, vol. 27, pp. 163–171.

Giordano, M.A., Giordano, M. and Wolf, A. 2002, 'The geography of water conflict and cooperation: Internal pressures and international manifestations', *Geographical Journal*, vol. 168, no. 4, pp. 293–312.

Giordano, M., Giordano, M.A. and Wolf, A.T. 2005, 'International resource conflict and mitigation', *Journal of Peace Research*, vol. 42, no. 1, pp. 47–65.

Giordano, M., Drieschova, A., Duncan, J., Sayama, Y., De Stefano, L. and Wolf, A. 2014, 'A review of the evolution and state of transboundary freshwater treaties', *International Environmental Agreements: Politics, Law and Economics*, vol. 14, no. 3, pp. 245–264.

Gleditsch, N.P., Brock, L., Homer-Dixon, T., Perelet, R. and Vlachos, E. (eds) 1997, *Conflict and the Environment*, Kluwer Academic Publishers, Dordrecht; Boston; London.

Gleditsch, N.P., Furlong, K., Hegre, H., Lacina, B. and Owen, T. 2006, 'Conflicts over shared rivers: Resource scarcity or fuzzy boundaries?', *Political Geography*, vol. 25, no. 4, pp. 361–382.

Gleick, P.H. 1993, 'Water and conflict: Fresh water resources and international security', *International Security*, vol. 18, no. 1, pp. 79–112.

Global Environmental Facility (GEF) 2012, *Contributing to Global Security: GEF Action on Water, Environment and Sustainable Livelihoods*, GEF, Washington, DC.

Global Water Partnership (GWP) 2000, *Integrated Water Resources Management*, Global Water Partnership, Stockholm, Sweden.

Government of India 2002, *Resolution No. 2/21/2002-BM*, Government of India, New Delhi.

Government of South Africa 1998, *SADC Launches Operation Boleas in Lesotho*. Available at: www.info.gov.za/speeches/1998/98a01_boleas9811173.htm.

Greacen, C.S. and Greacen, C. 2004, 'Thailand's electricity reforms: Privatization of benefits and socialization of costs and risks', *Pacific Affairs*, vol. 77, no. 3, The Political Economy of Electricity Reform in Asia, pp. 517–541.

Greacen, C. and Palettu, A. 2007, 'Electricity sector planning and hydropower in the Mekong Region', in *Democratizing Water Governance in the Mekong Region*, ed. L. Lebel, J. Dore, R. Daniel and Y.S. Koma, Silkworm Books, Chiang Mai, pp. 93–125.

Green Cross International 2000, *National Sovereignty and International Watercourses*, Green Cross International, Switzerland.

Griffin, R.S. 1991, *Recommendations for Strengthening the System and Working Relationships of the National Mekong Committees, National Planning and Implementing Agencies, and the Mekong Secretariat for Regional Cooperation: A Discussion Paper for the Mekong Secretariat*, Mekong Secretariat, Bangkok.

Grumbine, R.E. and Pandit, M.K. 2013, 'Threats from India's Himalaya dams', *Science*, vol. 339, no. 6115, pp. 36–37.

Gupta, J. and Lebel, L. 2010, 'Access and allocation in earth system governance: Water and climate change compared', *International Environmental Agreements: Politics, Law and Economics*, vol. 10, no. 4, pp. 377–395.

Guvi, C. 2008, 'SA keeps options open on Lesotho water project', *Business Day (South Africa)*, 27 September, p. 4.

Guzzini, S. 2005, 'The concept of power: A constructivist analysis', *Millennium – Journal of International Studies*, vol. 33, no. 3, pp. 495–521.

Gyawali, D. 2001, *Water in Nepal*, Himal Books and Panos South Asia with Nepal Water Conservation Foundation, Lalitpur.

Gyawali, D. and Dixit, A. 2000, 'Mahakali impasse: A futile paradigm's bequested travails', in *Domestic Conflict and Crisis of Governability in Nepal*, ed. D. Kumar, Centre for Nepal and Asian Studies, Kathmandu, pp. 236–304.

Haas, L., Mazzei, L. and O'Leary, D.T. 2010, *Lesotho Highlands Water Project: Communications Practices for Governance and Sustainabilty Improvement*, World Bank, Washington, DC.

Haddadin, M.J. 2001, 'Water scarcity impacts and potential conflicts in the MENA region', *Water International*, vol. 26, no. 4, pp. 460–470.

Haftendorn, H. 2000, 'Water and international conflict', *Third World Quarterly*, vol. 21, no. 1, pp. 51.

Handley, P. and Hiebert, M. 1992, 'Hostile undercurrents: Disputes deepen over use of Mekong River Water', *Far Eastern Economic Review*, 2 April.

Harris, L.M. 2002, 'Water and conflict geographies of the Southeastern Anatolia Project', *Society and Natural Resources*, vol. 15, no. 8, pp. 743–759.

Harris, L.M. and Alatout, S. 2010, 'Negotiating hydro-scales, forging states: Comparison of the upper Tigris/Euphrates and Jordan River basins', *Political Geography*, vol. 29, no. 3, pp. 148–156.

Hartmann, E. 2002, *Strategic Scarcity: Origins and Impacts of Environmental Conflict Ideas*, Ph.D. thesis, London School of Economics and Political Science, London.

Hayashi, K. 1999, *Nanbu Afurica Seiji Keizai Ron [Southern African Political Economy Analysis]*, Institute of Developing Economies, Chiba.

Heinmiller, T. 2009, 'Path dependency and collective action in common pool governance', *International Journal of the Commons*, vol. 3, no. 1, pp. 131–147.

Hendricks, L.B. 2006, 'Benefit sharing in transboundary waters: The SADC approach and the South African experience', Speech delivered by Mrs L. Hendricks, Minister of Water Affairs and Forestry at the Stockholm Water Symposium, Stockholm, Sweden, 21 August. Available at: www.info.gov.za/speeches/2006/06082210451001. htm.

—— 2008, Media Statement: Outcomes of Cabinet discussion on Water and Augmentation of the Vaal River System, Minister Lindiwe Hendricks, Department of Water Affairs and Forestry, 4 December, Government Communications (GCIS), Government of South Africa. Available at: www.info.gov.za/speeches/2008/08120412151002. htm.

Hensel, P.R., McLaughlin Mitchell, S. and Sowers II, T.E. 2006, 'Conflict management of riparian disputes', *Political Geography*, vol. 25, no. 4, pp. 383–411.

Hensengerth, O. 2009, 'Transboundary river cooperation and the regional public good: The case of the Mekong river', *Contemporary Southeast Asia: A Journal of International and Strategic Affairs*, vol. 31, no. 2, pp. 326–349.

Herbertson, K. 2013, *Xayaburi Dam: How Laos Violated the 1995 Mekong Agreement*, International Rivers, Berkeley, CA.

Heyns, P.S.V.H. 1995, 'The Namibian perspective on regional collaboration in the joint development of international water resources', *Water Resources Development*, vol. 11, no. 4, pp. 467–492.

—— 2003, 'Water-resources management in Southern Africa', in *International Waters in Southern Africa*, ed. M. Nakayama, United Nations University Press, Tokyo; New York.

Hiebert, M. 1991, 'The Mekong: The common stream; fertile imagination; muddy waters', *Far Eastern Economic Review*, vol. 151, no. 8, pp. 24–27.

Hirsch, P., Jensen, K.M., Boer, B., Carrard, N., FitzGerald, S. and Lyster, R. 2006, *National Interests and Transboundary Water Governance in the Mekong*, in collaboration with Danish International Development Assistance and the University of Sydney, Australian Mekong Resource Centre, Sydney.

Hisschemöller, M. and Hoppe, R. 2001, 'Coping with intractable controversies: The case for problem structuring in policy design and analysis', in *Knowledge, Power, and Participation in Environmental Policy Analysis*, ed. M. Hisschemöller, R. Hoppe, W.N. Dunn and J.R. Ravetz, *Policy Studies Review Annual*, Vol. 12, Transaction Publishers, New Brunswick, NJ, pp. 47–72.

Hitchcock, R.K. 2012, 'The Lesotho Highlands Water Project: Water, culture, and environmental change', in *Water, Cultural Diversity, and Global Environmental Change: Emerging Trends, Sustainable Futures?*, ed. B.R. Johnston, Springer; UNESCO, Dordrecht; Jakarta, pp. 319–338.

Homer-Dixon, T.F. 1991, 'On the threshold: Environmental changes as causes of acute conflict', *International Security*, vol. 16, no. 2, pp. 76–116.

—— 1994, 'Environmental scarcities and violent conflict: Evidence from cases', *International Security*, vol. 19, no. 1, pp. 5–40.

—— 1999, *Environment, Scarcity, and Violence*, Princeton University Press, Princeton, NJ.

Hooper, B.P. and Lloyd, G.J. 2011, *Report on IWRM in Transboundary Basins*, UNEP-DHI Centre for Water and Environment, Hørsholm.

Hoover, R. 2001, *Pipe Dreams: The World Bank's Failed Efforts to Restore Lives and Livelihoods of Dam-Affected People in Lesotho*, International Rivers Network, Berkeley, CA.

Hori, H. 1996, *Mekong-Gawa: Kaihatsu to Kankyou [The Mekong Rivers: Development and the Environment]*, Kokon-Shoin, Tokyo.
—— 2000, *The Mekong: Environment and Development*, United Nations University Press, Tokyo; New York.
Horta, K. 1995, 'The mountain kingdom's white oil: The Lesotho Highlands Water Project', *Ecologist*, vol. 25, no. 6, pp. 227–231.
Hossain, I. 1998, 'Bangladesh–India relations: The Ganges water-sharing treaty and beyond', *Asian Affairs: An American Review*, vol. 25, no. 3, pp. 131–150.
Huitema, D., Mostert, E., Egas, W., Moellenkamp, S., Pahl-Wostl, C. and Yalcin, R. 2009, 'Adaptive water governance; Assessing adaptive management from an institutional perspective', *Ecology and Society*, vol. 14, p. 26.
Huntjens, P., Lebel, L., Pahl-Wostl, C., Camkin, J., Schulze, R. and Kranz, N. 2012, 'Institutional design propositions for the governance of adaptation to climate change in the water sector', *Global Environmental Change*, vol. 22, no. 1, pp. 67–81.
Immerzeel, W.W., Pellicciotti, F. and Bierkens, M.F.P. 2013, 'Rising river flows throughout the twenty-first century in two Himalayan glacierized watersheds', *Nature Geoscience*, vol. 6, no. 9, pp. 742–745.
Intelligence Community, U. 2012, *Global Water Security: Intelligence Community Assessment*, Department of State, US, Washington, DC.
Interim Mekong Committee (IMC) 1991, *Draft Declaration of 4th November 1991 by The Members of the Mekong Committee (Committee for Co-ordination of investigations of the Lower Mekong Basin): Cambodia, Lao PDR, Thailand and Viet Nam*.
International Centre for Environmental Management (ICEM) 2010, *MRC Strategic Environmental Assessment (SEA) of Hydropower on the Mekong Mainstream*, ICEM, Hanoi.
International Rivers 2011, *Technical Review of the Xayaburi Environmental Impact Assessment*. Available at: www.internationalrivers.org/resources/technical-review-of-the-xayaburi-environmental-impact-assessment-3930.
Iyer, R.R. 1999, 'Conflict-resolution: Three river treaties', *Economic and Political Weekly*, vol. 34, no. 24, pp. 1509–1518.
Jacobs, I. 2012, 'A community in the Orange: The development of a multi-level water governance framework in the Orange–Senqu River basin in Southern Africa', *International Environmental Agreements: Politics, Law and Economics*, vol. 12, no. 2, pp. 187–210.
Jacobs, J.W., 1994, 'Toward sustainability in Lower Mekong River basin development', *Water International*, vol. 19, no. 1, pp. 43–51.
—— 1995, 'Mekong Committee history and lessons for river basin development', *The Geographical Journal*, vol. 161, no. 2, pp. 135–148.
—— 1999, 'Comparing river basin development experiences in the Mississippi and the Mekong', *Water International*, vol. 24, no. 3, pp. 196–203.
—— 2002, 'The Mekong River Commission: Transboundary water resources planning and regional security', *The Geographical Journal*, vol. 168, no. 4, pp. 354–364.
Jägerskog, A. 2003, *Why States Cooperate over Shared Water: The Water Negotiations in the Jordan River Basin*, Ph.D. thesis, Department of Water and Environmental Studies, Linköping University.
Jansky, L., Pachova, N.I. and Murakami, M. 2004, 'The Danube: A case study of sharing international waters', *Global Environmental Change*, vol. 14, pp. 39–49.
Japan International Cooperation Agency (JICA) 2001, *Mekon kasen ryuuiki suimon moderingu chousa: Jizen houkokusho [Mekong River Basin Hydro-meteorological Modelling Study: Pre-study Report]*, JICA, Tokyo.

Jaster, R.S. 1988, *The Defence of White Power: South African Foreign Policy under Pressure*, Macmillan Press in association with the International Institute for Strategic Studies, Basingstoke.

Jensen, K.M. and Lange, R.B. 2013, *Transboundary Water Governance in a Shifting Development Context*, Danish Institute for International Studies, Copenhagen.

Jeuland, M., Harshadeep, N., Escurra, J., Blackmore, D. and Sadoff, C. 2013, 'Implications of climate change for water resources development in the Ganges basin', *Water Policy*, vol. 15, no. S1, pp. 26–50.

Jha, B.K. 1973, *Indo–Nepalese Relations (1957–1972)*, Vora & Co. Publishers, Mumbai.

Jimenez, A. and Perez-Foguet, A. 2009, 'International investments in the water sector', *International Journal of Water Resources Development*, vol. 25, no. 1, pp. 1–14.

Joint Committee on Water Resources (JCWR) 2008, Minutes of the Third Meeting of the Nepal–India Joint Committee on Water Resources (JCWR) held from 29 September to 1 October, Kathmandu, Nepal.

—— 2009, Minutes of the Fourth Meeting of the India–Nepal Joint Committee on Water Resources (JCWR) held on 12–13 March, New Delhi, India.

Joint Declaration 1975, *Joint Declaration of Principles for Utilization of the Waters of the Lower Mekong Basin,* signed by the Representatives of the Government of Cambodia, Laos, Thailand and Vietnam to the Committee for Coordination of Investigations of the Lower Mekong Basin at Vientiane on 31 January.

Jordan, A. 2008, 'The governance of sustainable development: Taking stock and looking forwards', *Environment and Planning C: Government and Policy*, vol. 26, no. 1, pp. 17–33.

Kaika, M. 2006, 'Dams as symbols of modernization: The urbanization of nature between geographical imagination and materiality', *Annals of the Association of American Geographers*, vol. 96, no. 2, pp. 276–301.

Käkönen, M. 2008, 'Mekong Delta at the crossroads: More control or adaptation?', *Ambio*, vol. 37, no. 3, pp. 205–212.

Kathmandu Post 1999, 'Nepal–India joint committee urged to expedite work on disputed border', *Kathmandu Post*, 12 September.

Kay, A. 2005, 'A critique of the use of path dependency in policy studies', *Public Administration*, vol. 83, no. 3, pp. 553–571.

Keck, M.E. and Sikkink, K. 1998, *Activists Beyond Borders: Advocacy Networks in International Politics*, Cornell University Press, Ithaca, NY.

Keskinen, M., Mehtonen, K. and Varis, O. 2008, 'Transboundary cooperation vs. internal ambitions: The role of China and Cambodia in the Mekong region', in *International Water Security: Domestic Threats and Opportunities*, ed. N. Pachova, M. Nakayama and L. Jansky, United Nations University Press, Tokyo; New York, pp. 79–109.

Keskinen, M., Kummu, M., Käkönen, M. and Varis, O. 2012, 'Mekong at the crossroads: Next steps for impact assessment of large dams', *Ambio*, vol. 41, no. 3, pp. 319–324.

Khadka, N. 1988, 'Nepal's seventh Five-Year Plan', *Asian Survey*, vol. 28, no. 5, pp. 555–572.

Khilnani, S. 2003, *The Idea of India*, Penguin, London.

Kibaroğlu, A. 2002, *Building a Regime for the Waters of the Euphrates–Tigris River Basin*, Kluwer Law International, The Hague.

King Letsie III 2004, *Speech by His Majesty, King Letsie III*. Available at: www.lhwp.org.ls/news/apr04/speechbyking.htm.

Kingdom of Cambodia 2011, *Mekong River Commission Procedures for Notification, Prior Consultation and Agreement: Form/Format for Reply to Prior Consultation.* Available at: www.mrcmekong.org/news-and-events/consultations/xayaburi-hydropower-project-prior-consultation-process/.

Kingdom of Thailand 2011, *Mekong River Commission Procedures for Notification, Prior Consultation and Agreement: Form/Format for Reply to Prior Consultation.* Available at: www.mrcmekong.org/news-and-events/consultations/xayaburi-hydropower-project-prior-consultation-process/.

Kistin, E.J. 2007, 'Trans-boundary Water Cooperation in SADC: From Concept to Implementation', Paper prepared for the eighth WaterNet/WARFSA/GWP-SA Symposium, Lusaka, Zambia, 30 October to 3 November.

—— 2010, *Critiquing Cooperation: The Dynamic Effects of Transboundary Water Regimes*, D.Phil. thesis, Oxford University, Oxford.

Kistin, E.J. and Phillips, D. 2007, 'A Critique of Existing Agreements on Transboundary Waters, and Proposals for Creating Effective Cooperation between Co-riparians', Third International Workshop on Hydro-Hegemony, 12–13 May, London.

Kistin, E.J., Ashton, P.J., Earle, A., Malzbender, D., Patrick, M. and Turton, A.R. 2009, 'An overview of the content and historical context of the international freshwater agreements that South Africa has entered into with neighbouring countries', *International Environmental Agreements: Politics, Law and Economics*, vol. 9, no. 1, pp. 1–21.

Kistin Keller, E.J. 2012, 'Critiquing cooperation: Transboundary water governance and adaptive capacity in the Orange–Senqu Basin', *Journal of Contemporary Water Research and Education*, vol. 149, no. 1, pp. 41–55.

Kliot, N. 1993, *Water Resources and Conflict in the Middle East*, Routledge, London.

Klotz, A. and Lynch, C. 2007, *Strategies for Research in Constructivist International Relations: International Relations in a Constructed World*, M.E. Sharpe, Armonk, NY.

Komakech, H.C. and van der Zaag, P. 2011, 'Understanding the emergence and functioning of river committees in a catchment of the Pangani Basin Tanzania', *Water Alternatives*, vol. 4, no. 2, pp. 197–222.

Kooy, M. and Bakker, K. 2008, 'Technologies of government: Constituting subjectivities, spaces, and infrastructures in colonial and contemporary Jakarta', *International Journal of Urban and Regional Research*, vol. 32, no. 2, pp. 375–391.

Kosi Agreement 1954, *Agreement between His Majesty's Government of Nepal and the Government of India concerning the Kosi Project,* signed at Kathmandu on 25 April.

Kumar, D. 2004, 'Parliament and public policy making: A case study of the Mahakali Treaty', in *Nepal: Political Parties and Parliament*, ed. L.R. Baral, Adroit, Delhi, pp. 146–171.

Lahmeyer MacDonald Consortium and Olivier Shand Consortium 1986, *Lesotho Highlands Water Project. Feasibility Study: Final Report*, Ministry of Water, Energy and Mining, Lesotho, Maseru.

Lai, N.T. 2013, *Statement by H.E. Dr. Nguyen Thai Lai, 19th Meeting of the MRC Council.* Available at: www.mrcmekong.org/news-and-events/speeches/statement-by-h-e-dr-nguyen-thai-lai-19th-meeting-of-the-mrc-council/.

Lankford, B. and Hepworth, N. 2010, 'The cathedral and the bazaar: Monocentric and polycentric river basin management', *Water Alternatives*, vol. 3, no. 1, pp. 82–101.

Lao People's Democratic Republic (PDR) 2011, *Comments by Lao PDR on the MRCs*

Technical Review Report of the Proposed Xayaburi Dam Project. Available at: www.mrc-mekong.org/news-and-events/consultations/xayaburi-hydropower-project-prior-consultation-process/.

Laurence, P. 1984, 'Chances of Lesotho, SA security pact rise', *Rand Daily Mail*, 9 October, p. 4.

Laustsen, C.B. and Wæver, O. 2000, 'In defence of religion: Sacred referent objects for securitization', *Millennium – Journal of International Studies*, vol. 29, no. 3, pp. 705–739.

Lautze, J. and Giordano, M. 2007, 'Demanding supply management and supplying demand management: Transboundary waters in Sub-Saharan Africa', *The Journal of Environment and Development*, vol. 16, no. 3, pp. 290–306.

Lax, D.A. and Sebenius, J.K. 1986, *The Manager as Negotiator: Bargaining for Cooperation and Competitive Gain*, Free Press, New York; Collier Macmillan, London.

Le Billon, P. 2001, 'The political ecology of war: Natural resources and armed conflicts', *Political Geography*, vol. 20, no. 5, pp. 561–584.

Le Coq, J.F., Dufumier, M. and Trebuil, G. 2004, 'History of rice production in the Mekong Delta', in *Smallholders and Stockbreeders: Histories of Foodcrop and Livestock Farming in Southeast Asia*, ed. P. Boomgaard and D. Henley, KITLV Verhandelingen series, KITLV Press, Leiden, pp. 163–185.

Lebel, L., Grothmann, T. and Siebenhüner, B. 2010, 'The role of social learning in adaptiveness: Insights from water management', *International Environmental Agreements: Politics, Law and Economics*, vol. 10, no. 4, pp. 333–353.

Le-Huu, T. and Nguyen-Duc, L., 2003, *The Mekong. Case Study Prepared Under the UNESCO Water for Peace Programme 'From Potential Conflict to Cooperation Potential' (PCCP)*. UNESCO-IHP, Paris.

LeMarquand, D.G. 1977, *International Rivers: The Politics of Cooperation*, Westwater Research Centre, University of British Colombia, Vancouver, BC.

Lepono, T., Du Preez, H.H. and Thokoa, M. 2003, 'Monitoring of water transfer from Katse Dam into the Upper Vaal river system: Water utility's perspective', *Water Science and Technology*, vol. 48, no. 10, pp. 97–102.

Lesotho Highlands Development Authority (LHDA) n.d.a, *Lesotho Highlands Water Project 1986–1996*, LHDA, Maseru.

—— n.d.b, *Muela Hydropower Plant Project*, LHDA, Maseru.

Lesotho Highlands Development Authority–Trans-Caledon Tunnel Authority (LHDA–TCTA) 2001, *Partners for Life: Lesotho Highlands Water Project*, LHDA–TCTA, Maseru; Pretoria.

Lesotho Highlands Water Commission (LHWC) (Lesotho Delegation) 2004, 'Speech by the Minister of Natural Resources', *Lesotho Highlands Further Phases*, vol. 1, no. 1, pp. 1–2.

Lesotho Highlands Water Project (LHWP) 2010, *Water Sales*. Available at: www.lhwp.org.ls/Reports/PDF/Water%20Sales.pdf (accessed 18 August 2010).

Letter of Exchange 1996, *Letter of Exchange: Treaty Between His Majesty's Government of Nepal And The Government of India Concerning The Integrated Development of the Mahakali Barrage Including Sarada Barrage, Tanakpur Barrage and Pancheshwar Project*, signed in New Delhi on 12 February.

Likoti, F.J. 2007, 'The 1998 military intervention in Lesotho: SADC peace mission or resource war?', *International Peacekeeping*, vol. 14, no. 2, pp. 251–263.

Lipschutz, R.D. 1997, 'Environmental conflict and environmental determinism: The relative importance of social and natural factors', in *Conflict and the Environment*, ed.

N.P. Gleditsch, L. Brock, T. Homer-Dixon, R. Perelet and E. Vlachos, Kluwer Academic Publishers, Dordrecht; Boston, MA; London, pp. 35–50.

Lowi, M.R. 1993, *Water and Power: The Politics of a Scarce Resource in the Jordan River Basin*, Cambridge University Press, Cambridge; New York.

Mahakali Treaty 1996, *Treaty Between His Majesty's Government of Nepal And The Government of India Concerning The Integrated Development of the Mahakali Barrage Including Sarada Barrage, Tanakpur Barrage and Pancheshwar Project*, signed in New Delhi on 12 February.

Makim, A. 1997, *The Politics of Regional Cooperation on the Mekong: 1957–1995*, Ph.D. thesis, University of Queensland.

—— 2002, 'Resources for security and stability? The politics of regional cooperation on the Mekong, 1957–2001', *The Journal of Environment Development*, vol. 11, no. 5, pp. 5–52.

Martin, A., Rutagarama, E., Cascão, A., Gray, M. and Chhotray, V. 2011, 'Understanding the co-existence of conflict and cooperation: Transboundary ecosystem management in the Virunga Massif', *Journal of Peace Research*, vol. 48, no. 5, pp. 621–635.

Marty, F. 2001, *Managing International Rivers: Problems, Politics and Institutions*, Peter Lang, Bern.

Mason, M. and Zeitoun, M. 2013, 'Questioning environmental security', *The Geographical Journal*, vol. 179, no. 4, pp. 294–297.

Mathews, J.T. 1989, 'Redefining security', *Foreign Affairs*, vol. 68, no. 2, pp. 162–177.

Matlosa, K.T. 2000, 'The Lesotho Highlands Water Project: Socio-economic impacts', in *Environmental Security in Southern Africa*, ed. D. Tevera and S. Moyo, Southern Africa Regional Institute for Policy Studies, SAPES Trust, Harare.

Matthew, R. 1997, 'Rethinking environmental security', in *Conflict and the Environment*, ed. N. Gleditsch, Springer, Netherlands, pp. 71–90.

—— 2013, 'Climate change and water security in the Himalayan region', *Asia Policy*, vol. 16, no. 1, pp. 39–44.

Matthews, N. 2012, 'Water grabbing in the Mekong Basin: An analysis of the winners and losers of Thailand's hydropower development in Lao PDR', *Water Alternatives*, vol. 5, no. 2, pp. 392–411.

Mbeki, T. 2004, *Speech by President of the Republic of South Africa, Mr. Thabo Mbeki*. Available at: www.lhwp.org.ls/news/apr04/speechbypresident.htm.

McMillan, S.M. 1997, 'Interdependence and conflict', *Mershon International Studies Review*, vol. 41, no. 1, pp. 33–58.

Mehta, L. 2003, 'Contexts and constructions of water scarcity', *Economic and Political Weekly*, vol. 38, no. 48, pp. 5066–5072.

—— 2010a, 'Afterword: Looking beyond scarcity?', in *The Limits to Scarcity: Contesting the Politics of Allocation*, ed. L. Mehta, Earthscan Publications, London, pp. 253–256.

—— 2010b, 'Commentary', in *The Limits to Scarcity: Contesting the Politics of Allocation*, ed. L. Mehta, Earthscan Publications, London, pp. 145–147.

—— 2010c, 'The scare, naturalization and politicization of scarcity', in *The Limits to Scarcity: Contesting the Politics of Allocation*, ed. L. Mehta, Earthscan Publications, London, pp. 13–30.

Meissner, R. and Turton, A.R. 2003, 'The hydrosocial contract theory and the Lesotho Highlands Water Project', *Water Policy*, vol. 5, no. 2, pp. 115–126.

Mekong Committee (MC) 1964, Committee for Co-ordination of Investigations of the Lower Mekong Basin: Twenty-Third Session (Plenary), 7–13 January, MC, Bangkok, Thailand.

—— 1975, Report of the Sixty-Eighth Session (Plenary) of the Mekong Committee, 29 January to 3 February, MC, Vientiane Laos, Bangkok, Thailand.

Mekong River Commission (MRC) 2003a, Minutes of the Tenth Meeting of the Council, Mekong River Commission, 29–30 November, Phnom Penh, Cambodia, MRC, Phnom Penh.

—— 2003b, *Procedures for Notification, Prior Consultation and Agreement*, MRC, Phnom Penh.

—— 2007, *Independent Organisational, Financial and Institutional Review of the Mekong River Commission, Secretariat and the National Mekong Committees*, MRC, Vientiane.

—— 2010, *State of the Basin Report 2010*, MRC, Vientiane.

—— 2013, *Mekong Basin Planning: The Story Behind the Basin Development Plan*, MRC, Vientiane.

Mekong Secretariat 1977a, *Thailand and the Mekong Project*, Mekong Secretariat, Bangkok, Thailand.

—— 1977b, *Viet-Nam and the Mekong Project*, Bangkok: Mekong Secretariat, Bangkok, Thailand.

—— 1989, *The Mekong Committee: A Historical Account (1957–89)*, Bangkok: Mekong Secretariat, Bangkok, Thailand.

Mekonnen, D.Z. 2010, 'The Nile Basin Cooperative Framework Agreement negotiations and the adoption of a "Water Security" paradigm: Flight into obscurity or a logical cul-de-sac?', *European Journal of International Law*, vol. 21, no. 2, pp. 421–440.

Menga, F. 2014, *Power and Dams in Central Asia*, Ph.D. thesis, The University of Cagliari, Cagliari.

Menniken, T. 2008, *Hydrological Regionalism in the Mekong and Nile Basin*, Ph.D. thesis, Albert-Ludwigs-Universität, Freiburg.

Menon, P.K. 1972, 'Some institutional aspects of the Mekong Basin Development Committee', *International Review of Administrative Sciences*, vol. 38, pp. 157–168.

Merme, V., Ahlers, R. and Gupta, J. 2014, 'Private equity, public affair: Hydropower financing in the Mekong Basin', *Global Environmental Change*, vol. 24, pp. 20–29.

Merrey, D.J. 2009, 'African models for transnational river basin organisations in Africa: An unexplored dimension', *Water Alternatives*, vol. 2, no. 2, pp. 183–204.

Middleton, C., Garcia, J. and Foran, T. 2009, 'Old and new hydropower players in the Mekong Region: Agendas and strategies', in *Contested Waterscapes in the Mekong Region: Hydropower, Livelihoods and Governance*, ed. F. Molle, T. Foran and M. Käkönen, Earthscan, London, pp. 23–54.

Middleton, C., Matthews, N. and Mirumachi, N. 2014, 'Whose risky business?: Public private partnerships and large hydropower dams in the Mekong Region', in *Hydropower Development in the Mekong Region: Political, Socio-economic and Environmental Perspectives*, ed. N. Matthews and K. Geheb, Routledge, London.

Mills, G. 1992, 'Lesotho: Between independence and incorporation', in *Southern Africa at the Crossroads? Prospects for Stability and Development in the 1990s*, ed. L. Benjamin and C. Gregory, Justified Press, Rivonia, pp. 63–73.

Milman, A. and Scott, C.A. 2010, 'Beneath the surface: Intranational institutions and management of the United States–Mexico transboundary Santa Cruz aquifer', *Environment and Planning C: Government and Policy*, vol. 28, no. 3, pp. 528–551.

Milner, H. 1992, 'International theories of cooperation among nations: Strengths and weaknesses', *World Politics*, vol. 44, no. 3, pp. 466–496.

Mirumachi, N. 2007a, 'Fluxing relations in water history: Conceptualizing the range of relations in transboundary river basins', *Pasts and Futures of Water: Proceedings from the 5th International Water History Association Conference*, 13–17 June, Tampere, Finland.

—— 2007b, 'Introducing Transboundary Waters Interaction NexuS (TWINS): Model of interaction dynamics in transboundary waters', *Third International Workshop on Hydro-Hegemony*, 12–13 May, London.

—— 2008, 'Domestic issues in developing international waters in Lesotho: Ensuring water security amidst political instability', in *International Water Security: Domestic Threats and Opportunities*, ed. N. Pachova, M. Nakayama and L. Jansky, United Nations University Press, Tokyo; New York, pp. 35–60.

—— 2012, 'How domestic water policies influence international transboundary water development: A case study of Thailand', in *Politics and Development in a Transboundary Watershed – The Case of the Lower Mekong Basin*, ed. J. Öjendal, S. Hansson and S. Hellberg, Springer Verlag, Heidelberg, pp. 83–100.

—— 2013, 'Transboundary water security: Reviewing the importance of national regulatory and accountability capacities', in *Water Security: Principles, Perspectives and Practices*, ed. B. Lankford, K. Bakker, M. Zeitoun and D. Conway, Routledge, London, pp. 166–179.

—— 2015, 'Water and sustainable development', in *Routledge International Handbook of Sustainable Development*, ed. M. Redclift and D Springett, Routledge, London, pp. 136–146.

Mirumachi, N. and Allan, J.A. 2007, 'Revisiting transboundary water governance: Power, conflict, cooperation and the political economy', *International Conference on Adaptive and Integrated Water Management (CD-ROM)*, 12–15 November, Basel, Switzerland.

Mirumachi, N. and Torriti, J. 2012, 'The use of public participation and economic appraisal for public involvement in large-scale hydropower projects: Case study of the Nam Theun 2 Hydropower Project', *Energy Policy*, vol. 47, pp. 125–132.

Mirumachi, N. and van Wyk, E. 2010, 'Cooperation at different scales: Challenges for local and international water resource governance in South Africa', *Geographical Journal*, vol. 176, no. 1, pp. 25–38.

Mirumachi, N. and Warner, J. 2008, 'Co-existing Conflict and Cooperation in Transboundary Waters', Paper presented at the International Studies Association Annual Conference, 26–30 March, San Francisco, CA.

Mirumachi, N., Zeitoun, M. and Warner, J. 2013, 'Transboundary water interaction and the 1997 UN Watercourses Convention: Allocation of waters and implementation of principles', in *The UN Watercourses Convention in Force: Strengthening International Law for Transboundary Water Management*, ed. A. Rieu-Clarke and F. Loures, Routledge, London, pp. 352–364.

Mitchell, S.M. and Hensel, P.R. 2007, 'International institutions and compliance with agreements', *American Journal of Political Science*, vol. 51, no. 4, pp. 721–737.

Mitrany, D. 1975, *The Functional Theory of Politics*, London School of Economics and Political Science, London.

Mizra, M.M.Q. 2004, 'Hydrological changes in Bangladesh', in *The Ganges Water Diversion: Environmental Effects and Implications*, ed. M.M.Q. Mizra, Kluwer Academic Publishers, Dordrecht, pp. 13–38.

Mochizuki, K. 1998, 'Lesoto ga komutta nann: Minshuka no shougeki' [Lesotho's crisis: Impacts of democratization], in *Minami Afurica no Shougeki: Posuto Mandela ki no Seijikeizai [The Impacts of South Africa: Political Economy of the Post-Mandela Period]*, ed. K. Hirano, Institute of Developing Economies, Chiba, pp. 127–143.

Mohamed, A.E. 2003, 'Joint development and cooperation in international water resources', in *International Waters in Southern Africa*, ed. M. Nakayama, United Nations University Press, Tokyo, pp. 209–248.

Mokorosi, P.S. and van der Zaag, P. 2007, 'Can local people also gain from benefit sharing in water resources development? Experiences from dam development in the Orange–Senqu River Basin', *Physics and Chemistry of the Earth, Parts A/B/C*, vol. 32, nos 15–18, pp. 1322–1329.

Molle, F. 2002, 'The Closure of the Chao Phraya River Basin in Thailand: Its Causes, Consequences and Policy Implications', Paper presented at the Conference on Asian Irrigation in Transition – Responding to the Challenges Ahead, Workshop, 22–23 April, Asian Institute of Technology, Bangkok, Thailand.

—— 2007, 'Scales and power in river basin management: The Chao Phraya River in Thailand', *Geographical Journal*, vol. 173, no. 4, pp. 358–373.

—— 2008, 'Why enough is never enough: The societal determinants of river basin closure', *International Journal of Water Resources Development*, vol. 24, no. 2, pp. 217–226.

Molle, F. and Hoanh, C.T. 2009, *Implementing Integrated River Basin Management: Lessons from the Red River Basin, Vietnam*, IWMI, Colombo.

Molle, F., Mollinga, P.P. and Wester, P. 2009, 'Hydraulic bureaucracies and the hydraulic mission: Flows of water, flows of power', *Water Alternatives*, vol. 2, no. 3, pp. 328–349.

Mollinga, P.P. 2008a, 'Water, politics and development: Framing a political sociology of water resource management', *Water Alternatives*, vol. 1, no. 1, pp. 7–23.

—— 2008b, 'Water policy – Water politics: Social engineering and strategic action in water sector reform', in *Water Politics and Development Cooperation: Local Power Plays and Global Governance*, ed. W. Scheumann, S. Neubert and M. Kipping, Springer-Verlag, Heidelberg, pp. 1–29.

—— 2010, 'The material conditions of a polarized discourse: Clamours and silences in critical analysis of agricultural water use in India', *Journal of Agrarian Change*, vol. 10, no. 3, pp. 414–436.

Murthy, P. 1999, 'India and Nepal: Security and economic dimensions', *Strategic Analysis*, vol. 23, no. 9, pp. 1531–1547.

Mustafa, D. 2007, 'Social construction of hydropolitics: The geographical scales of water and security in the Indus Basin', *Geographical Review*, vol. 97, no. 4, pp. 484–501.

Myers, N. 1989, 'Environment and security', *Foreign Policy*, no. 74, pp. 23–41.

Myint, T. 2012, *Governing International Rivers: Polycentric Politics in the Mekong and the Rhine*, Edward Elgar, Cheltenham.

Nakayama, M. 1999, 'Aspects behind differences in two agreements adopted by riparian countries of the Lower Mekong River Basin', *Journal of Comparative Policy Analysis: Research and Practice*, vol. 1, no. 3, pp. 293–308.

Narayanamoorthy, A. 2005, *Where Water Seeps!: Towards a New Phase in India's Irrigation Reforms*, Academic Foundation, New Delhi.

Narayanan, N.C. and Venot, J. 2009, 'Drivers of change in fragile environments: Challenges to governance in Indian wetlands', *Natural Resources Forum*, vol. 33, no. 4, pp. 320–333.

Nation, The 1991, 'Phnom Penh attends Mekong meeting', *The Nation*, 5 November.
—— 1977, 'A major breakthrough for Mekong Project', *The Nation*, 30 April, p. 1.
National Planning Commission (Government of Nepal) 1981, *The Sixth Plan (1980–1985): A Summary*, Government of Nepal, Kathmandu.
—— 1984, *Basic Principles of the Seventh Plan (1985–1990)*, Government of Nepal, Kathmandu.
Neumann, I.B. 1998, 'Identity and the outbreak of war: Or why the Copenhagen School of security studies should include the idea of "violisation" in its framework of analysis", *International Journal of Peace Studies*, vol. 3, no. 1, pp. 1–10.
Newig, J., Pahl-Wostl, C. and Sigel, K. 2005, 'The role of public participation in managing uncertainty in the implementation of the Water Framework Directive', *European Environment*, vol. 15, no. 6, pp. 333–343.
Nicol, A., van Steenbergen, F., Sunman, H., Turton, A.R., Slaymaker, T., Allan, J.A., de Graaf, M. and van Harten, M. 2001, *Transboundary Water Management as an International Public Good: Prepared for The Ministry of Foreign Affairs, Sweden*, Swedish Ministry of Foreign Affairs, Stockholm.
Nilsson, C., Reidy, C.A., Dynesius, M. and Revenga, C. 2005, 'Fragmentation and flow regulation of the world's large river systems', *Science*, vol. 308, no. 5720, pp. 405–408.
Ninham Shand Consulting Engineers n.d., *The Lesotho Highlands Water Project, Inauguration of Phase 1A: The Role of Ninham Shand Consulting Engineers*, Ninham Shand Consulting Engineers, Cape Town.
Nthako, S. and Griffiths, A.L. 1997, 'Lesotho Highlands water project – Project management', *Proceedings of the Institution of Civil Engineers. Civil Engineering*, vol. 120, no. 1, pp. 3–13.
O'Neill, K., Balsiger, J. and VanDeveer, S.D. 2004, 'Actors, norms, and impact: Recent international cooperation theory and the influence of the agent-structure debate', *Annual Review of Political Science*, vol. 7, no. 1, pp. 149–175.
Ohlsson, L. 1999, *Environment, Scarcity and Conflict: A Study of Malthusian Concerns*, Ph.D. thesis, Department of Peace and Development Research, University of Göteborg, Sweden.
—— 2000, 'Water conflicts and social resource scarcity', *Physics and Chemistry of the Earth, Part B: Hydrology, Oceans and Atmosphere*, vol. 25, no. 3, pp. 213–220.
Öjendal, J. 2000, *Sharing the Good: Modes of Managing Water Resources in the Lower Mekong River Basin*, Department of Peace and Development Research, University of Göteborg, Sweden.
Orange–Senqu River Awareness Kit 2010, *Water Infrastructure: Dams and Infrastructure: Lesotho Highlands Water Project*. Available at: www.orangesenqurak.org/challenge/infrastructure/dams+and+infrastructure/lhwp.aspx (accessed 18 August 2010).
Orange–Senqu River Commission (ORASECOM) 2000, *Agreement between the Governments of the Republic of Botswana, the Kingdom of Lesotho, the Republic of Namibia and the Republic of South Africa on the Establishment of the Orange–Senqu River Commission*.
—— 2013a, *The Orange–Senqu River Basin: Infrastructure Catalogue*, 2nd edn, ORASECOM, Pretoria.
—— 2013b, *Transboundary Environmental Assessment in the Orange–Senqu River Basin: Recommendations of the ORASECOM Council*, ORASECOM, Pretoria.
—— 2014, *Orange–Senqu River Basin: Transboundary Diagnostic Analysis*, ORASECOM, Pretoria.

Ostrom, E. 1990, *Governing the Commons: The Evolution of Institutions for Collective Action*, Cambridge University Press, Cambridge; New York.

—— 1999, 'Coping with tragedies of the commons', *Annual Review of Political Science*, vol. 2, no. 1, pp. 493–535.

—— 2010, 'Beyond markets and states: Polycentric governance of complex economic systems', *The American Economic Review*, vol. 100, no. 3, pp. 641–672.

Oye, K.A. 1985, 'Explaining cooperation under anarchy: Hypotheses and strategies', *World Politics*, vol. 38, no. 1, pp. 1–24.

Pahl-Wostl, C. 2009, 'A conceptual framework for analysing adaptive capacity and multi-level learning processes in resource governance regimes', *Global Environmental Change*, vol. 19, no. 3, pp. 354–365.

Pearce, F. 2012, 'China is taking control of Asia's water tower', *New Scientist*, vol. 2862.

Pherudi, M.L. 2003, 'Lesotho: Political conflict, peace and reconciliation in the mountain kingdom', in *Through Fire with Water: The Roots of Division and the Potential for Reconciliation in Africa*, ed. E. Doxtader and C. Villa-Vicencio, Africa World Press, Trenton, NJ.

Phillips, D., Daoudy, M., McCaffrey, S.C., Öjendal, J. and Turton, A.R. 2006, *Trans-boundary Water Cooperation as a Tool for Conflict Prevention and for Broader Benefit-sharing: Prepared for the Ministry for Foreign Affairs, Sweden*, EGDI Secretariat, Ministry for Foreign Affairs, Stockholm.

Pingali, P.L., Tri Khiem, N., Gerpacio, R.V. and Xuan, V. 1997, 'Prospects for sustaining Vietnam's reacquired rice exporter status', *Food Policy*, vol. 22, no. 4, pp. 345–358.

Planning Commission (Government of India) 1952, *First Five Year Plan*, Government of India, New Delhi.

—— 1960, *Third Five Year Plan*, Government of India, New Delhi.

—— 2002, *Tenth Five Year Plan*, Government of India, New Delhi.

Pokharel, J.C. 1991, *Environmental Resource Negotiation Between Asymmetrically Powerful Nations: Power of the Weaker Nations*, Ph.D. thesis, Massachusetts Institute of Technology.

Pomeranz, K. 2013, 'Asia's unstable water tower: The politics, economics, and ecology of Himalayan water projects', *Asia Policy*, vol. 16, no. 1, pp. 4–10.

Pun, S.B. 2009, 'Power trading', in *The Nepal–India Water Relationship: Challenges*, ed. D.N. Dhungel and S.B. Pun, Springer, Dordrecht, pp. 153–196.

Radosevich, G.E. 1995, *Agreement on the Cooperation for the Sustainable Development of the Mekong River Basin: Commentary and History*, UNDP, Bangkok, Thailand.

—— 1996, 'The Mekong – A new framework for development and management', in *Asian International Waters: From Ganges-Brahmaputra to Mekong*, ed. A.K. Biswas and T. Hashimoto, Oxford University Press, Bombay; New York, pp. 245–278.

Ramoeli, P. 2002, 'The SADC Protocol on Shared Watercourses: Its origins and current status', in *Hydropolitics in the Developing World: A Southern African Perspective*, ed. A.R. Turton and R. Henwood, African Water Issues Research Unit (AWIRU), University of Pretoria, Pretoria, pp. 105–112.

Rand Daily Mail 1984, 'Pik threatens to end Lesotho water venture', *Rand Daily Mail*, 27 August, p. 2.

Räsänen, T.A., Koponen, J., Lauri, H. and Kummu, M. 2012, 'Downstream hydrological impacts of hydropower development in the Upper Mekong Basin', *Water Resources Management*, vol. 26, pp. 3495–3513.

Raskin, P., Gleick, P., Kirshen, P., Pontius, G. and Strzepek, K. 1997, *Water Futures: Assessment of Long-range Patterns and Prospects*, Stockholm Environment Institute, Stockholm.

Rasul, G. 2014, 'Food, water, and energy security in South Asia: A nexus perspective from the Hindu Kush Himalayan region', *Environmental Science and Policy*, vol. 39, pp. 35–48.

Reber, P. 1998, 'Lesotho king, Mandela open regional water project', *Associated Press*, 22 January.

Reisner, M. 1993, *Cadillac Desert: The American West and its Disappearing Water*, revised and updated edn, Penguin Books, New York.

Renner, M. 1996, *Fighting for Survival: Environmental Decline, Social Conflict, and the New Age of Insecurity*, W.W. Norton, New York.

Rieu-Clarke, A., Moynihan, R. and Magsig, B. 2012, *UN Watercourses Convention: User's Guide*, IHP-HELP Centre for Water Law, Policy and Science (under the auspices of UNESCO), Dundee.

Ringius, L. 2001, *Radioactive Waste Disposal at Sea: Public Ideas, Transnational Policy Entrepreneurs, and Environmental Regimes*, MIT Press, Cambridge, MA.

Roberts, E. and Finnegan, L. 2013, *Building Peace around Water, Land and Food: Policy and Practice for Preventing Conflict*, Quaker United Nations Office, Geneva.

Rockström, J., Steffen, W., Noone, K., Persson, Å., Chapin, I.,F.S., Lambin, E., Lenton, T.M., Scheffer, M., Folke, C., Schellnhuber, H., Nykvist, B., De Wit, C.A., Hughes, T., van der Leeuw, S., Rodhe, H., Sörlin, S., Snyder, P.K., Costanza, R., Svedin, U., Falkenmark, M., Karlberg, L., Corell, R.W., Fabry, V.J., Hansen, J., Walker, B., Liverman, D., Richardson, K., Crutzen, P. and Foley, J. 2009, 'Planetary boundaries: Exploring the safe operating space for humanity', *Ecology and Society*, vol. 14, no. 2, 32pp.

Roe, P. 2006, 'Reconstructing identities or managing minorities? Desecuritizing minority rights: A response to Jutila', *Security Dialogue*, vol. 37, no. 3, pp. 425–438.

Royal Academy of Engineering 2010, *Global Water Security: An Engineering Perspective*, The Royal Academy of Engineering, London.

Sadoff, C.W. and Grey, D. 2002, 'Beyond the river: The benefits of cooperation on international rivers', *Water Policy*, vol. 4, pp. 389–403.

—— 2005, 'Cooperation on international rivers: A continuum for securing and sharing benefits', *Water International*, vol. 30, no. 4, pp. 420–427.

Salgado, I. 1998, 'Mandela turns on taps for mammoth water scheme in S. Africa, Lesotho', *Agence France Presse*, 22 January.

Salman, S.M.A. and Uprety, K. 2002, *Conflict and Cooperation on South Asia's International Rivers: A Legal Perspective*, World Bank, Washington, DC.

Sandler, T. 2006, 'Regional public goods and regional cooperation', in *Expert Paper Series Seven: Cross-Cutting Issues*, ed. Secretariat of the International Task Force on Global Public Goods, Secretariat of the International Task Force on Global Public Goods, Stockholm, pp. 143–178.

Sangchai, S., 1967, *The Mekong Committee: A New Genus of International Organization*, unpublished Ph.D. thesis, Indiana University.

SAPA News Agency 1999, 'South African minister holds talks with Lesotho premier on water treaty', *SAPA News Agency*, 23 April.

Schmeier, S. 2013, *Governing International Watercourses: River Basin Organizations and the Sustainable Governance of Internationally Shared Rivers and Lakes*, Routledge, London.

Schnurr, M. 2008, 'Global water governance: Managing complexity on a global scale', in *Water Politics and Development Cooperation*, ed. W. Scheumann, S. Neubert and M. Kipping, Springer, Heidelberg, pp. 107–120.

Schwartzstein, P. 2013, 'Water wars: Egyptians condemn Ethiopia's Nile dam project', *National Geographic*, 27 September. Available at: http://news.nationalgeographic.com/news/2013/09/130927-grand-ethiopian-renaissance-dam-egypt-water-wars/.

Scudder, T. 2005, *The Future of Large Dams: Dealing with Social, Environmental and Political Costs*, Earthscan, London.

—— 2006, 'Assessing the impacts of the LHWP on resettled households and other affected people 1986–2005', in *On the Wrong Side of Development: Lessons Learned from the Lesotho Highlands Water Project*, ed. M.L. Thamae and L. Pottinger, Transformation Resource Centre, Maseru, pp. 39–87.

Searle, J.R. 1969, *Speech Acts: An Essay in the Philosophy of Language*, Cambridge University Press, Cambridge.

Sebastian, A.G. 2008, *Transboundary Water Politics: Conflict, Cooperation, and Shadows of the Past in the Okavango and Orange River Basins of Southern Africa*, Ph.D. thesis, University of Maryland, College Park.

Sebenius, J.K. 1992, 'Negotiation analysis: A characterization and review', *Management Science*, vol. 38, no. 1, pp. 18–38.

Sebesvari, Z., Le, Huang Thi Thu Le, Toan, P.V., Arnold, U. and Renaud, F.G. 2012, 'Agriculture and water quality in the Vietnamese Mekong Delta', in *The Mekong Delta System: Interdisciplinary Analyses of a River Delta*, ed. F.G. Renaud and C. Kuenzer, Springer, Berlin, pp. 331–362.

Seckler, D., Amarasinghe, U., Molden, D.J., de Silva, R. and Barker, R. 1998, *World Water Demand and Supply, 1990 to 2025: Scenarios and Issues*, IWMI, Colombo.

Secretary for Water Affairs (Government of South Africa) 1971, *Third Supplementary Report on the First Phase of the Orange River Development Project: Prepared in Terms of Section 58 of the Water Act, 1958*, DWA, Pretoria.

Sehring, J. 2009. 'Path dependencies and institutional bricolage in post-Soviet water governance', *Water Alternatives*, vol. 2, no. 1, pp. 61–81.

Seid, A.H., Fekade, W. and Olet, E. 2013, 'The Nile Basin Initiative: Advancing transboundary cooperation and supporting riparian communities', in *Free Flow: Reaching Water Security through Cooperation*, ed. UNESCO, UNESCO, Paris, pp. 35–38.

Selby, J. 2003, 'Dressing up domination as "cooperation": The case of the Israeli–Palestinian water relations', *Review of International Studies*, vol. 29, pp. 121–138.

Sewell, W.R.D. and White, G.F. 1966, *The Lower Mekong: An Experiment in International River Development*, International Conciliation No. 558, Carnegie Endowment for International Peace, New York.

Shah, T. and Giordano, M. 2013, 'Himalayan water security: A South Asian perspective', *Asia Policy*, vol. 16, no. 1, pp. 26–31.

Shapland, G. 1997, *Rivers of Discord: International Water Disputes in the Middle East*, Hurst & Company, London.

Shrestha, H.M. and Singh, L.M. 1996, 'The Ganges–Brahmaputra system: A Nepalese perspective in the context of regional cooperation', in *Asian International Waters: From Ganges-Brahmaputra to Mekong*, ed. A.K. Biswa and T. Hashimoto, Oxford University Press, Bombay; New York, pp. 81–94.

Shuman, E. 2001, *Israel Warns Lebanon Over Hatzbani Water Project. Israel Insider*, 15

March. Available at: www.israelinsider.com/channels/diplomacy/articles/dip_0008. html.

Simons, H.J. 1968, 'Harnessing the Orange River', in *Dams in Africa: An Interdisciplinary Study of Man-Made Lakes in Africa*, ed. N. Rubin and W.M. Warren, Frank Cass, London, pp. 128–145.

Simpson, A. 2007, 'The environment–energy security nexus: Critical analysis of an energy "love triangle" in Southeast Asia', *Third World Quarterly*, vol. 28, no. 3, pp. 539–554.

Smith, L.D. and Porter, K. 2010, 'Management of catchments for the protection of water resources: Drawing on the New York City watershed experience', *Regional Environmental Change*, vol. 10, no. 4, pp. 311–326.

Sneddon, C. 2003, 'Reconfiguring scale and power: The Khong-Chi-Mun Project in Northeast Thailand', *Environment and Planning A*, vol. 35, no. 12, pp. 2229–2250.

—— 2013, 'Water, governance and hegemony', in *Contemporary Water Governance in the Global South: Scarcity, Marketization and Participation*, ed. L.M. Harris, J.A. Goldin and C. Sneddon, Routledge, London, pp. 13–24.

Sneddon, C. and Fox, C. 2006, 'Rethinking transboundary waters: A critical hydropolitics of the Mekong basin', *Political Geography*, vol. 25, no. 2, pp. 181–202.

—— 2012, 'Water, geopolitics, and economic development in the conceptualization of a region', *Eurasian Geography and Economics*, vol. 35, no. 1, pp. 143–160.

Socialist Republic of Viet Nam 2011, *Mekong River Commission Procedures for Notification, Prior Consultation and Agreement: Form/Format for Reply to Prior Consultation*. Available at: www.mrcmekong.org/news-and-events/consultations/xayaburi-hydropower-project-prior-consultation-process/.

Sojamo, S. 2008, 'Illustrating co-existing conflict and cooperation in the Aral Sea Basin with TWINS approach', in *Central Asian Waters: Social, Economic, Environmental and Governance Puzzle*, ed. M.M. Rahaman and O. Varis, Water & Development Publications, Helsinki University of Technology, Espoo, pp. 75–88.

Soroos, M.S. 1994, 'Global change, environmental security, and the prisoner's dilemma', *Journal of Peace Research*, vol. 31, no. 3, pp. 317–332.

Sosland, J.K. 2007, *Cooperating Rivals: The Riparian Politics of the Jordan River Basin*, State University of New York Press, New York.

Southern African Development Community (SADC) and European Development Fund (EDF) 2009, *Proposals for Stakeholder Participation in ORASECOM*, SADC and EDF, Gaborone.

Southern African Development Community (SADC) 2010, *SADC's Regional Project on Economic Accounting of Water Use: Project Report*, SADC, Gaborone.

Statute of MC 1957, *Statute of the Committee for Co-ordination of Investigations of Lower Mekong Basin Established by the Governments of Cambodia, Laos, Thailand and the Other Republics of Viet-Nam in Response to the Decision Taken by the United Nations Economic Commission for Asia and the Far East*, Phonm-Pehn (Cambodia), 31 October.

Strategic Foresight Group 2011, *Himalayan Solutions: Co-operation and Security in River Basins*, Strategic Foresight Group, Mumbai.

Stritzel, H. 2007, 'Towards a theory of securitization: Copenhagen and beyond', *European Journal of International Relations*, vol. 13, no. 3, pp. 357–383.

Suhardiman, D. and Giordano, M. 2012, 'Process-focused analysis in transboundary water governance research', *International Environmental Agreements: Politics, Law and Economics*, vol. 12, no. 3, pp. 299–308.

Suhardiman, D., Giordano, M. and Molle, F. 2012, 'Scalar disconnect: The logic of transboundary water governance in the Mekong', *Society and Natural Resources*, vol. 25, no. 6, pp. 572–586.

Sultana, F. 2004, 'Engendering a catastrophe: A gendered analysis of India's River-Linking Project', in *Regional Cooperation on Transboundary Rivers: Impacts of the Indian River-Linking Project*, ed. M.F. Ahmed, Q.K. Ahmad and M. Khalequzzman, BAPA Press, Dhaka, pp. 288–305.

Swatuk, L.A. 2008, 'A political economy of water in Southern Africa', *Water Alternatives*, vol. 1, no. 1, pp. 24–47.

Swatuk, L. and Wirkus, L. 2009, 'Transboundary water governance in southern Africa: An introduction', in *Transboundary Water Governance in Southern Africa: Examining Underexplored Dimensions*, ed. L. Swatuk and L. Wirkus, Nomos Press, Baden-Baden, pp. 11–29.

Swyngedouw, E. 2004, 'Spain's hydraulic mission: Conflict, power, and mastering of water', *International Symposium on Water Resources Management: Risks and Challenges for the 21st Century 2–4 September 2004, Izmir, Turkey*, ed. N.B. Harmancioglu, O. Fistikoglu, Y. Dalkilic and A. Gul, pp. 117–130.

Takahashi, K., 1974, *Framework for Multinational Regional Development: A Case Study in the International Administrative and Financial Cooperation in the Program to Develop the Lower Mekong Basin*, unpublished Ph.D. thesis, New York University.

Tekateka, R. 2011, 'Transboundary water management issues under the NWA and regional collaboration, policies and conventions', in *Transforming Water Management in South Africa: Designing and Implementing a New Policy Framework*, ed. B. Schreiner and R. Hassan, Springer, New York; Heidelberg, pp. 253–270.

Thai Mekong Community Network of 8 Provinces 2011, *Statement 20 April 2011: Demand for Lao Government to Halt Xayaburi Dam Construction to Avoid Conflict with Mekong Countries*. Available at: www.livingriversiam.org/4river-tran/4mk/mek_ne191.html.

Tingsanchali, T. and Singh, P.R. 1996, 'Optimum water resources allocation for Mekong-Chi-Mun Transbasin Irrigation Project, Northeast Thailand', *Water International*, vol. 21, no. 1, pp. 20–29.

Toset, H.P.W., Gleditsch, N.P. and Hegre, H. 2000, 'Shared rivers and interstate conflict', *Political Geography*, vol. 19, pp. 971–996.

Trans-Caledon Tunnel Authority – Lesotho Highlands Development Authority (TCTA–LHDA) 2003, *Sustainable Development: Lesotho Highlands Water Project*, TCTA–LHDA, Pretoria.

Treaty on the LWHP 1986, *Treaty on the Lesotho Highlands Water Project between the Government of the Kingdom of Lesotho and the Government of the Republic of South Africa*, signed in Maseru, 24 October.

Trombetta, M.J. 2011, 'Rethinking the securitization of environment: Old beliefs, new insights', in *Securitization Theory: How Security Problems Emerge and Dissolve*, ed. T. Balzacq, Routledge, London, pp. 135–149.

Tromp, L. 2006, 'Lesotho Highlands: The socio-economics of exporting water', *Proceedings of ICE – Civil Engineering*, vol. 159, pp. 44–49.

Trottier, J. 2003, *Water Wars: The Rise of a Hegemonic Concept – Exploring the Making of the Water War and Water Peace Belief within the Israeli–Palestinian Conflict*, UNESCO IHP, Paris.

Trudeau, H., Duplessis, I., Lalonde, S., Van de Graaf, T., De Ville, F., O'Neill, K., Roger, C., Dauvergne, P., Morin, J., Oberthür, S., Orsini, A., Biermann, F., Ohta,

H. and Ishii, A. 2013, 'Insights from global environmental governance', *International Studies Review*, vol. 15, no. 4, pp. 562–589.

Tu, D.T. 2002, 'Land and water investment in Viet Nam: Past trends, returns and future requirements', in *Investment in Land and Water. Proceedings of the Regional Consultation, 3–5 October 2001, Bangkok, Thailand*, FAO, Bangkok, Thailand, pp. 338–349.

Tuomela, R. 2000, *Cooperation: A Philosophical Study*, Kluwer Academic Publishers, Dordrecht; Boston, MA.

Turton, A.R. 2010, *New Thinking on the Governance of Water and River Basins in Africa: Lessons from the SADC Region*, South African Institute of International Affairs (SAIIA), Johannesburg.

Turton, A.R. and Funke, N. 2008, 'Hydro-hegemony in the context of the Orange River Basin', *Water Policy*, vol. 10, Supplement 2, pp. 51–69.

Turton, A.R. and Meissner, R. 2002, 'The hydrosocial contract and its manifestation in society: A South African case study', in *Hydropolitics in the Developing World: A Southern African Perspective*, ed. A.R. Turton and R. Henwood, African Water Issues Research Unit (AWIRU), University of Pretoria, Pretoria, pp. 37–60.

Turton, A.R., Meissner, R., Mampane, P.M. and Seremo, O. 2004, *A Hydropolitical History of South Africa's International River Basins*, WRC, Gezina.

Turton, A.R., Schultz, C., Buckle, H., Kgomongoe, M., Malungani, T. and Drackner, M. 2006, 'Gold, scorched earth and water: The hydropolitics of Johannesburg', *Water Resources Development*, vol. 22, no. 2, pp. 313–335.

Tvedt, T. 2004, *The River Nile in the Age of the British: Political Ecology and the Quest for Economic Power*, I.B. Tauris, London.

Ullman, R.H. 1983, 'Redefining security', *International Security*, vol. 8, no. 1, pp. 129–153.

UN Water 2013a, *Water Cooperation in Action: Approaches, Tools and Processes – Conference Report of the International Annual UN-Water Zaragoza Conference 2012/2013; Preparing for the 2013 International Year – Water Cooperation: Making it Happen*, UN Water Decade Programme on Advocacy and Communication, Zaragoza.

—— 2013b, *Water Security and the Global Agenda: A UN-Water Analytical Brief*, UNU University, Hamilton, Ontario.

United Nations Development Programme (UNDP) 2006, *Human Development Report 2006: Beyond Scarcity: Power, Poverty and the Global Water Crisis*, Palgrave Macmillan, Basingstoke; New York.

Untawale, M.G. 1974, 'The political dynamics of functional collaboration: Indo–Nepalese River Projects', *Asian Survey*, vol. 14, no. 8, pp. 716–732.

Upadhyay, S.K. 1991, *Tryst with Diplomacy*, Vikas, New Delhi.

van der Zaag, P. and Carmo Vaz, Á. 2003, 'Sharing the Incomati waters: Cooperation and competition in the balance', *Water Policy*, vol. 5, pp. 349–368.

van der Zaag, P. and Gupta, J. 2008, 'Scale issues in the governance of water storage projects', *Water Resources Research*, vol. 44, no. 10, pp. W10417.

Vasquez, J.A. 1995, 'Why global conflict resolution is possible: Meeting the challenges of the new world order', in *Beyond Confrontation: Learning Conflict Resolution in the Post-Cold War Era*, ed. J.A. Vasquez, J.T. Johnson, S.M. Jaffe and L. Stamato, University of Michigan Press, Ann Arbor, pp. 131–154.

Venot, J., Bharati, L., Giordano, M. and Molle, F. 2011, 'Beyond water, beyond boundaries: Spaces of water management in the Krishna river basin, South India', *The Geographical Journal*, vol. 177, no. 2, pp. 160–170.

Verghese, B.G. 1999, *Waters of Hope: From Vision to Reality in Himalaya–Ganga Development Cooperation*, 2nd edn, Oxford & IBH, New Delhi.

Verghese, B.G., Iyer, R.R. and Centre for Policy Research 1993, *Harnessing the Eastern Himalayan Rivers: Regional Cooperation in South Asia*, Konark Publishers, New Delhi.

Vörösmarty, C.J., McIntyre, P.B., Gessner, M.O., Dudgeon, D., Prusevich, A., Green, P., Glidden, S., Bunn, S.E., Sullivan, C.A., Liermann, C.R. and Davies, P.M. 2010, 'Global threats to human water security and river biodiversity', *Nature*, vol. 467, no. 7315, pp. 555–561.

Wade, R. 1988, *Village Republics: Economic Conditions for Collective Action in South India*, Cambridge University Press, Cambridge; New York.

Wæver, O. 1995, 'Securitization and desecuritization', in *On Security*, ed. R.D. Lipschutz, Columbia University Press, New York, pp. 46–86.

Wallis, S. 1992, *Lesotho Highlands Water Project Volume 1*, Reprinted 1994 edn, Laserline, Surrey.

—— 1995, *Lesotho Highlands Water Project Volume 3*, Laserline, Surrey.

Warner, J. 2004a, *Plugging the GAP – Working with Buzan: The Ilisu Dam as a Security Issue*, SOAS/King's College London, London.

—— 2004b, 'Water, wine, vinegar, blood: On politics, participation, violence and conflict over the hydrosocial contract', *Proceedings of the Workshop on Water and Politics: Understanding the Role of Politics in Water Management, Marseille, February 26–27, 2004*, ed. World Water Council, World Water Council, pp. 7–18.

—— 2005, 'Multi-stakeholder platforms: Integrating society in water resource management?', *Ambiente & Sociedade*, vol. 8, no. 2, pp. 4–28.

—— 2008, 'Contested hydrohegemony: Hydraulic control and security in Turkey', *Water Alternatives*, vol. 1, no. 2, pp. 271–288.

—— 2011, *Flood Planning: The Politics of Water Security*, I.B. Tauris, London.

Warner, J. and van Buuren, A. 2009, 'Multi-stakeholder learning and fighting on the River Scheldt', *International Negotiation*, vol. 14, no. 2, pp. 419–440.

Warner, J. and Wegerich, K. 2010, 'Is water politics? Towards international water relations', in *Politics of Water: A Survey*, ed. K. Wegerich and J. Warner, Taylor & Francis, London.

Warner, J.F. and Zeitoun, M. 2008, 'International relations theory and water do mix: A response to Furlong's troubled waters, hydro-hegemony and international water relations', *Political Geography*, vol. 27, no. 7, pp. 802–810.

Warner, J., Wester, P. and Bolding, A. 2008, 'Going with the flow: River basins as the natural units for water management?', *Water Policy*, vol. 10, no. S2, pp. 121–138.

Waterbury, J. 2002, *The Nile Basin: National Determinants of Collective Action*, Yale University Press, New Haven, CT.

Watershed 2003, 'Use of the Mekong water agreed (sort of)', *Watershed*, vol. 9, no. 2, p. 6.

Wegerich, K. and Warner, J. 2010, *Politics of Water: A Survey*, Routledge, London.

Wegerich, K., Kazbekov, J., Mukhamedova, N. and Musayev, S. 2012, 'Is it possible to shift to hydrological boundaries? The Ferghana Valley Meshed System', *International Journal of Water Resources Development*, vol. 28, no. 3, pp. 545–564.

Weinthal, E. 2002, *State Making and Environmental Cooperation: Linking Domestic and International Politics in Central Asia*, MIT Press, Cambridge, MA.

Wendt, A. 1999, *Social Theory of International Politics*, Cambridge University Press, Cambridge; New York.

Wessels, J. 2007, *To Cooperate or not to Cooperate...? Collective Action for Rehabilitation of Traditional Water Tunnel Systems (Qanats) in Syria*, Ph.D. thesis, University of Amsterdam, Amsterdam.

Wester, P. 2008, *Shedding the Waters: Institutional Change and Water Control in the Lerma-Chapala Basin, Mexico*, Ph.D. thesis, Wageningen University, Wageningen.

Westing, A.H. (ed.) 1986, *Global Resources and International Conflict: Environmental Factors in Strategic Policy and Action*, Oxford University Press, Oxford: New York.

White, G.F. 1957, 'A perspective of river basin development', *Law and Contemporary Problems*, vol. 22, no. 2, pp. 157–187.

Whittington, D., Wu, X. and Sadoff, C.W. 2005, 'Water resources management in the Nile basin: The economic value of cooperation', *Water Policy*, vol. 7, no. 3, pp. 227–252.

Wirsing, R.G. and Jasparro, C. 2007, 'River rivalry: Water disputes, resource insecurity and diplomatic deadlock in South Asia', *Water Policy*, vol. 9, no. 3, pp. 231–251.

Wisuttisak, P. 2012, 'Regulation and competition issues in Thai electricity sector', *Energy Policy*, vol. 44, pp. 185–198.

Wolf, A.T. 1995, *Hydropolitics along the Jordan River: Scarce Water and its Impact on the Arab–Israeli Conflict*, United Nations University Press, Tokyo; New York.

—— 1998, 'Conflict and cooperation along international waterways', *Water Policy*, vol. 1, no. 2, pp. 251–265.

—— 2007, 'Shared waters: Conflict and cooperation', *Annual Review of Environment and Resources*, vol. 32, no. 1, pp. 241–269.

Wolf, A.T., Yoffe, S.B. and Giordano, M. 2003, 'International waters: Identifying basins at risk', *Water Policy*, vol. 5, no. 1, pp. 29–60.

Wong, C.M., Williams, C.E., Pittock, J., Collier, U. and Schelle, P. 2007, *World's Top 10 Rivers at Risk*, WWF International, Gland, Switzerland.

Wong, S.M.T. 2010, *Making the Mekong: Nature, Region, Postcoloniality*, Ohio State University.

World Bank 2010a, *South Asia Water Initiative (SAWI) Multi-Donor Trust Fund: Annual Report FY10 (July 2009–June 2010) Prepared by the World Bank for the 3rd Annual Donors Meeting*. Available at: www.southasiawaterinitiative.org/.

—— 2010b, *World Development Report Development and Climate Change*, World Bank, Washington, DC.

—— 2012, *Ganges Strategic Basin Assessment: A Discussion of Regional Opportunities and Risks*, World Bank, Washington, DC.

World Commission on Dams (WCD) Secretariat 2000, *Orange River Development Project, South Africa: Case Study Prepared as an Input to the World Commission on Dams*, World Commission on Dams, Cape Town.

Worrell, S. 2012a, 'Water ministers urge Laos to halt Xayaburi', *The Phnom Penh Post*, 2 May.

—— 2012b, 'Cambodia, Vietnam united on Xayaburi', *The Phnom Penh Post*, 5 July.

Wu, X. and Whittington, D. 2006, 'Incentive compatibility and conflict resolution in international river basins: A case study of the Nile Basin', *Water Resources Research*, vol. 42, W02417.

Wuilbercq, E. 2014, 'Ethiopia's Nile dam project signals its intention to become an African power', *Guardian Weekly*, 14 July. Available at: www.theguardian.com/global-development/2014/jul/14/ethiopia-grand-renaissance-dam-egypt.

World Water Assessment Programme (WWAP) 2003, *1st UN World Water Development Report: Water for People, Water for Life*, UNESCO and Berghahn Books, Paris; New York; Oxford.

—— 2009, *The United Nations World Water Development Report 3: Water in a Changing World*, UNESCO; Earthscan, Paris; London.

—— 2012, *The United Nations World Water Development Report 4: Managing Water under Uncertainty and Risk*, UNESCO, Paris.

World Wide Fund for Nature (WWF) 2014, *A Request to Suspend or Cancel the Power Purchase Agreement from Xayaburi Dam*. Available at: awsassets.panda.org/downloads/xayaburi_dam_letter_english_final.pdf.

Wyatt, A.B. and Baird, I.G. 2007, 'Transboundary impact assessment in the Sesan River Basin: The case of the Yali Falls Dam', *International Journal of Water Resources Development*, vol. 23, no. 3, pp. 427–442.

Xinhua General News Service 1992, 'Nepalese parties to protest against government', *Xinhua General News Service*, 7 September.

—— 1993a, 'Ruling Nepalese party rejects opposition's demand of PM resignation', *Xinhua General News Service*, 7 February.

—— 1993b, 'Strike staged in Kathmandu valley', *Xinhua General News Service*, 14 March.

Yin, R.K. 2009, *Case Study Research: Design and Methods*, 4th edn, Sage, Los Angeles, CA.

Yoffe, S.B., Wolf, A.T. and Giordano, M. 2003, 'Conflict and cooperation over international freshwater resources: Indicators of basins at risk', *Journal of the American Water Resources Association*, vol. 39, no. 5, pp. 1109–1126.

Yong, M.L. and Grundy-Warr, C. 2012, 'Tangled nets of discourse and turbines of development: Lower Mekong mainstream dam debates', *Third World Quarterly*, vol. 33, no. 6, pp. 1037–1058.

Zawahri, N.A. 2008, 'Capturing the nature of cooperation, unstable cooperation, and conflict over international rivers: The story of the Indus, Yarmouk, Euphrates, and Tigris Rivers', *International Journal of Global Environmental Issues*, vol. 8, no. 3, pp. 286–310.

Zawahri, N. and Gerlak, A. 2009, 'Navigating international river disputes to avert conflict', *International Negotiation*, vol. 14, no. 2, pp. 211–227.

Zeitoun, M. 2006, *Power and the Palestinian–Israeli Water Conflict: Towards an Analytical Framework of Hydro-Hegemony*, Ph.D. thesis, King's College London, London.

Zeitoun, M. and Allan, J.A. 2008, 'Applying hegemony and power theory to trans-boundary water analysis', *Water Policy*, vol. 10, Supplement 2, pp. 3–12.

Zeitoun, M. and Mirumachi, N. 2008, 'Transboundary water interaction I: Reconsidering conflict and cooperation', *International Environmental Agreements: Politics, Law and Economics*, vol. 8, no. 4, pp. 297–316.

Zeitoun, M. and Warner, J. 2006, 'Hydro-hegemony: A framework for analysis of trans-boundary water conflicts', *Water Policy*, vol. 8, no. 5, pp. 435–460.

Zeitoun, M., Mirumachi, N. and Warner, J. 2011, 'Transboundary water interaction II: Soft power underlying conflict and cooperation', *International Environmental Agreements: Politics, Law and Economics*, vol. 11, pp. 159–178.

Zwarteveen, M.Z. 1997, 'Water: From basic need to commodity: A discussion on gender and water rights in the context of irrigation', *World Development*, vol. 25, no. 8, pp. 1335–1349.

Index

Page numbers in **bold** denote figures.